White Sky

A Year In Saudi Arabia

Mike Trial

Published by AKA:yoLa
Columbia Missouri
www.akayola.com

Library of Congress
Trial, Mike: White Sky, A Year in Saudi Arabia

ISBN-13: 978-0-9842288-1-2
ISBN-10: 0-9842288-1-0

Also available in Hardcover:
ISBN-13: 978-0-9842288-5-0
ISBN-10: 0-9842288-5-3

All names used in this memoir are fictional.

AKA:yoLa

Foreword

In the northwest corner of the Saudi Arabian peninsula, the desert is a sand and gravel plain between mountain ridges bare of foliage. Dust devils meander toward the mountains like wandering djinns in the blast-furnace heat. The mountains are bare rock, but if you look closely, you can see pale green and purple and red minerals lacing the rock.

Even on clear days, the sky is a hazy white from the ever-present dust. When sandstorms sweep in from the south, the sky turns a murky orange, and visibility drops to a few meters. These shamals, as the Arabs call them, sometimes last for days. While the wind is blowing, the bedouins stay in their tents, their ghutrahs wrapped around mouths and noses to keep the dust out.

Since antiquity people have lived and traded here, their culture shaped by the desert, and by the trade routes that converge here, and by conquest. Over the centuries, empires have risen and fallen, changes have come and gone. Today, change is coming again to the desert. Not far from the ancient trade town of Tabuk, the Saudis are investing hundreds of millions of their oil dollars to build a fully self-contained military base with all associated infrastructure. To complete this enterprise, the Saudi government requested assistance from the American government. The U.S. government assigned the task to the State Department's International Infrastructure Assistance Agency, IIAA for short.

IIAA's people changed the desert, but in ways both obvious and subtle, the desert's white heat forged greater changes in the people who came to live and work there.

Chapter 1

The Airbus A320 slipped down through the humid night toward a landing at Jidda, Saudi Arabia. Mark Exner watched a brilliantly lit tent city gleaming in white floodlights slide by, then the runway lights appeared, and the plane settled to a flawless landing. He sat for a moment, quiet in a sea of movement.

Two days ago he and Jennifer had been sitting on the benches in the park on the University of Missouri campus. "I love Columbia in autumn," Mark said. They soaked up the sun, plastic cups of Cabernet Sauvignon at hand. The breeze pushed a strand of Jennifer's long black hair over her sunglasses and she brushed it away with a familiar gesture.

"I'll only be gone a year," he said softly, as much to himself as to her. "I hate to leave, but if I let this opportunity go by, I'll always regret it."

The next day she drove him to the St. Louis airport. They said little, avoiding each others eyes. At the airline counter, he felt Jennifer's hand touch his – a warm, small touch. They sat together at the departure gate saying nothing until the flight was called.

Tears leaked out of her eyes. He gave her a long hug, joined the line at the boarding ramp, looked back once as he ducked inside, then made his way to his seat.

The hours passed as he flew from St. Louis to New York, then out over the midnight Atlantic. In the darkened plane, Mark sat staring at the seat back in front of him wondering what was ahead. He'd been looking forward to this move for months, but now...

Eventually he fell into a troubled sleep as the plane droned eastward in the white moonlit clouds above the black Atlantic. At the London airport Mark changed a hundred American dollars for four hundred Saudi riyals, then boarded his flight for Jidda.

He was exhausted, but sleep was the farthest thing from his mind. Saudi Arabia! He couldn't imagine what it would be like. Five years ago he'd graduated from the College of Engineering at the University of Missouri. He'd gotten lots of job offers, engineers were in demand, and after evaluating the offers, he'd chosen to go to work for the International Infrastructure Assistance Agency, IIAA, in their Washington D.C. office. The reason he'd taken that position, at a lower salary than he could have gotten at several of the commercial firms, was that they had the biggest, most challenging projects around. It had been a hectic five years in Washington, learning the practical aspects of design work to match the theoretical knowledge he'd gained in school. IIAA's design work was done in the States, but the projects themselves were constructed overseas, so now Mark wanted to take the next step, to travel overseas, and take part in the construction of those projects. He'd signed a one-year agreement with IIAA, to come to Saudi Arabia to live and work.

The plane lumbered toward the terminal. Announcements in Arabic and English went unheard as passengers swayed and jostled in the aisle of the still-taxiing plane, pulling their luggage out of the overhead compartments.

Mark stayed in his seat surreptitiously studying the small card IIAA had given him before he left home. It was a set of supposedly helpful phrases in Arabic, and an entirely incomprehensible map of Jidda. An IIAA employee was to meet Mark at the airport, but if that did not happen, he could show the card to a taxi driver and be taken to the IIAA office.

Mark hoped he'd be met; his few riyals wouldn't go far in this expensive city, and the IIAA office would be closed this time of night. He slipped the card back into his pocket.

He let the crowd push ahead of him out into the sultry night air, then followed them across the tarmac to the terminal.

Inside, it was freezer-cold. Passengers filed past two Saudis wearing blue uniforms with white helmet liners who looked to be about fourteen years old except for their hard dark eyes. In the baggage claim area, Mark stood at the back of the crowd. If there was no one to meet him, maybe he could just sit in the airport until morning.

The carousel began to turn and the crowd jammed closer; suitcases were bumped off the carousel.

Collecting his bags, he followed the crowd to customs, opened them both so an unsmiling Saudi kid in white thobe and red-and-white ghutrah could look through them. When that was done, he started for the doorway when another Saudi guard gestured him back. The man reached over and put a yellow chalk mark on the end of each suitcase, then turned back to his conversation as though Mark no longer existed.

Mark went through the doorway and around the corner. In the next room was a painted steel barricade and beyond that, filling the space shoulder to shoulder all the way to the glass doors, was a huge crowd of Sudanese, Yemenis, Thais, Lebanese, Egyptians, Palestinians, Koreans, every nationality, all talking at once. The noise was deafening.

Mark slowed, scanning the crowd for an American face, but found none. He got behind two Korean businessmen as they entered the crowd.

"Mark Exner?" a voice bellowed nearby. A huge man, bearded, two hundred fifty pounds, in a faded blue pullover shirt with a tiny IIAA name tag erupted from the pandemonium and thrust his hand toward Mark. They struggled to shake hands in the shoving, elbowing crowd. The big man's sign with Mark's name on it had been crushed into origami.

"I didn't see your sign," Mark shouted.

The big man grinned. "Follow me."

He hoisted one of Mark's suitcases over his head, bulled his way through the crowd and out into the humid night. Bumper-to-bumper traffic rumbled past on the street. In the glare of neon at the front of the arrivals terminal, a sandy-haired woman in a long blue jelabah stood at the curb next to an idling Chevy Suburban parked part way up on the sidewalk. A Filipino driver wearing a faded tee shirt with an IIAA logo was waving cars past. Horns blared continuously.

The big man threw Mark's suitcase in the Chevy and they scrambled in. The driver pulled out into traffic, inches from the oncoming cars, and accelerated across two lanes.

The big man grinned at Mark, "I'm Ron Stevens and this is my wife Laura."

"Nice to meet you. I'm Mark Exner." Mark breathed a great sigh of relief. "Boy, am I glad to see you."

Outside the window, traffic, store lights, and faces flowed by in a blur.

Saudi Arabia – he was here.

They drove through heavy traffic, all new cars, under bright sodium halide street lights. Shops were gaudy with neon. None of the streets were straight and none of the intersections were at right angles to each other. Traffic lights were installed but not working so every intersection became an eight lane game of chicken, cars slowing just enough to gauge each other's intent, then accelerating, missing each other by inches.

The driver turned down an unlighted alley. "Should have the traffic lights working soon," Ron said. The Suburban jounced through foot-deep potholes.

Laura laughed, "Most drivers don't pay attention to traffic lights even when they are working."

The Suburban stopped at a gate in a wall. There was no marking of any kind on it. They piled out, pushed through the unlocked gate and into the house, which turned out to be quite modern inside.

"This is a house IIAA is leasing. We use it and a couple of others as temporary quarters for our folks in Jidda for a few days. Very few hotels yet, and they're all super-expensive. All foreign companies rent houses like these."

They clomped up concrete steps to one of several bedrooms. "Everything should be all set," Ron said, glancing around. There was carpet on the floor, each room had a bed neatly made. "There are towels and such in the closet. Maid service comes in every few days."

"And the kitchen is fully furnished with plates, silverware, pots and pans," Laura added.

"But you won't be cooking anything," Ron said. "I'll come by to get you in the morning. We'll eat breakfast at IIAA's cafeteria." He pulled a couple of plastic bottles of water out of a box marked 'Sohat'.

"Never drink water out of the faucet; only drink bottled water," he said sternly, then grinned.

He shook hands with Mark. "Welcome to Saudi Arabia. I'll come by at say, eight o'clock?"

Mark nodded and they filed out.

Mark was suddenly so exhausted he could barely move.

He placed his suitcase on the bed and began to unpack. It seemed unreal, unpacking this same familiar suitcase which he had just packed a day ago in the morning light of Missouri. He stared at the folded shirts for a moment, put thoughts of Columbia and Jennifer out of his mind, and stretched out with a paperback to read for a few moments.

The light wasn't too good, but he could see well

enough to read. He tried to sink into Hemingway's *The Sun Also Rises* but his mind was restless.

He heard a knock at the front door downstairs.

Ron Stevens was standing there, grinning sheepishly. "I forgot to ask you…would you like to come over for a drink?"

The Stevens' house was one of twelve inside a gated compound. The house was the same dun-colored stucco as all the others. The Chevy's headlights spotlighted it once, bottom to top, as the Suburban went over another set of the ubiquitous speed bumps. Persian carpets covered black and white marble floor tiles. New American style furniture was arranged around the bare walls.

"I have scotch, vodka, Drambuie, some kind of gin," Ron called from the kitchen.

"Scotch would be fine. No ice."

"We're not really drinkers," Laura said. She had pageboy cut sandy hair, a profile from a Greek sculpture, and a ready smile.

"Alcohol is illegal in Saudi Arabia isn't it?" Mark said, to make conversation.

Laura laughed "American government employees are given special dispensation. We call it a 'tea ration'. Each employee is allowed to buy four bottles a month from the IIAA commissary. All very secret but…"

"… a completely transparent fiction," Ron finished. "The Saudis know all about it. But secrecy saves their face and ours. Johnny Walker okay?"

"Sure, thanks." Mark accepted the glass. "Do all American's get this special treatment?"

9

"Not the employees of private companies, no. And that sometimes causes resentment. But American government employees do, the American embassy, USMTM..."

"USMTM?"

"U.S. Military Training Mission." Ron sank into a chair that creaked ominously under his weight. "U.S. Army warrant officers training Saudi troops in everything from military tactics to tank maintenance."

The Scotch rose in Mark's blood. "Must be quite a task. I assume many of the troops are bedouins, just in from the desert."

"Some, yeah, but don't underestimate the Saudis. All the Saudi military officers, and the government officials, of which there are many, and the businessmen, are highly educated, very literate and well-travelled. More so than most of the Americans you'll meet here."

"And quite young for the level of responsibility they have," Laura added.

"You'll meet Saudi army lieutenants in charge of hundreds of millions of dollars of construction. Don't let their youth and relatively junior rank fool you. The Kingdom is basically using the army to simultaneously build a defense force, build a civil infrastructure, and train a civilian workforce in basic industrial practice. It's an ambitious undertaking."

"They are also using this tactic to distribute the oil income into the general economy so it doesn't concentrate in the hands of a few rich families, and they are succeeding at that too," Laura added. "But enough lecture, tell us about yourself. You're from

Missouri, aren't you?"

"Originally, yeah, but I've been working at the IIAA office In Washington D.C. for the last five years."

"Whatever induced you to come here?" Barbara laughed. "Do you have a mysterious past you're trying to get away from, like people joining the French Foreign Legion?"

Mark laughed, "No, just looking for challenging work and a chance to travel. Plus it sounded like a good chance for promotions..."

"It is," Ron said. "You won't make the big salaries the private company guys here get, but you'll get lots of challenge, lots of responsibility. IIAA manages some big projects, a lot bigger than you or I would be working on back in the States." He waved his hand around the room. "And we live pretty comfortably too."

"Not as good as it is for the guys working for private companies," Ron said. "They don't have to pay American income tax after they've been overseas for a year. We pay the same income tax we would in the States."

"Is all this American tax money?"

"None of it. Not even our salaries. It's Saudi money paid to the US Treasury and reimbursed to IIAA. The Saudis are paying for everything, the new construction, our salaries, our leased houses, cars, airfare, everything. In turn, they get reliable management since we are a U.S. government agency. They've been burned before by private firms that promise the moon and don't deliver. And the Saudis don't have enough people to effectively manage all the hundreds of projects themselves. They're trying to jump into the twentieth century in a single jump. I admire their ambition."

"One of the best features of working here is the opportunity to travel on your vacations." Laura added. "Go to Egypt, see the pyramids, go to Africa, see the animals, go to Europe…"

The doorbell rang. Ron went to the door and let in a young American dressed in jeans, western shirt, cowboy boots. "Mark, this is Alan Mack from personnel services."

"Nice to meet you Mark," the young guy said. "I'm glad I caught you tonight. We're changing your assignment. You're being reassigned to the Tabuk office. We need you up there tomorrow."

Mark stared at Alan.

"The mechanical engineer up in Tabuk just quit. They radioed me this afternoon. Guy just packed his bags, drove to the airport in Amman, probably on a flight to the States by now." Alan shook his head. "That reminds me. I've got to get somebody to go up to the Amman airport and bring our car back." Alan handed Mark a set of folded papers, "Here's your reassignment." Alan left as quickly as he'd come.

With a smile Laura topped off Mark's glass. "This kind of stuff happens from time to time. People just get fed up and leave."

"Not to worry, Mark." Ron said. He settled back into his chair. "I know the guy who quit. He wasn't happy from the day he got here. Some people like it here. Some people hate it."

After Ron drove him back to his transient quarters, Mark dropped into an exhausted sleep, dreaming he was adrift on a great midnight ocean. Dark shapes moved below him in ominous silence. Waking with a start, he got up and strolled over to the window. Nothing moved on the street below. Three a.m. and he was wide awake – jet lag.

He read until the coming dawn turned the room grey, then he got up and tried to open the cheap aluminum frame window. It slid open an inch and stuck. He pressed his nose up to the opening, inhaling the air of the dusty dawn. He ran a comb through his hair, then went out to the quiet street. The air was calm, smelling of dust and diesel and other scents he couldn't name. The still air was exactly the temperature of his skin.

Concrete block houses in desert colors, each behind its wall lined the street. Mark pulled the wrought iron gate closed behind him and strolled to the corner. In front of a store with its corrugated shutter pulled down and locked, cases of canned Pepsi were stacked. The familiar red, white and blue cans were lettered in English on one side, Arabic on the other. After a while, he wandered back to his room to wait for Ron.

Later that morning Mark boarded a small turboprop plane leased by IIAA. Two hours later the plane descended into the white desert haze and made a straight-in approach to a new asphalt airstrip. The barren red mountains in the distance shimmered in

13

the furnace-like heat. By the time they finished taxiing, Mark's shirt back was soaked with sweat. He followed the other passengers across the pavement to the terminal building through heat and glare that gave the landscape a strange underwater feeling. Although people were talking, it seemed very silent.

Inside the tiny terminal, at least it was cool. Mark lined up with the others waiting to show passports to a young Saudi in red and white checked ghutrah and white thobe. After the guard took a cursory look at his passport, Mark collected his suitcases from a handcart and stood in the shade looking past the chain link fence to the desert and the mountains beyond. Except for a dust devil far out across the gravel plain, nothing moved. In front of the terminal building, there were several Chevy Suburbans parked in the glaring sun. One was idling.

A hawk-nosed American with thinning red hair and aviator sunglasses got out and sauntered up to Mark.

"You Mark?"

"Yeah."

The man stuck out a bony hand and gave Mark a single pump of a handshake. "I'm Andy Petri, the executive officer here at the IIAA Tabuk office." He picked up one of Mark's suitcases. "How much more do you have?"

"This is it."

"That's a pleasant change. Some new people bring in ten, fifteen, suitcases full of crap."

They climbed into the white Chevy Suburban with air conditioning blasting and drove down an asphalt road, coming to an untended gate in a chain link fence. They went through and, as the Suburban jounced over interminable speed bumps, Ron pointed to various buildings, his descriptions incomprehensible. "That's the HSC hospital." He pointed to a three story building, obviously new.

"HSC?"

"It's run for the Saudi Defense Forces by the Health Services Corporation out of Dallas. HSC for short."

"American?"

"American company, but the staff is mostly Brits and Filipinos. Americans are too expensive."

Mark saw a couple of young western women in hospital whites walking in the front door of the hospital. Toyota pickup trucks and Chevy Suburbans were parked haphazardly around the unpaved parking lot.

"Plus a bunch of Jordanians, Egyptians, and Pakistanis. The usual mix," Andy added with disdain.

"This is the IIAA compound," he said as they drove through another untended gate. They drove past various concrete block administrative buildings. Mark felt like he was on some 1950's West Texas army base.

They passed a row of small dusty houses made out of fiber board, obviously temporary buildings long past their design lifespan. In front and back of the houses were rows of tamarisk trees with irrigation ditches. Some of the trees were forty feet tall.

"IIAA plant these trees?" Mark asked. "They must grow really fast."

"Nah. These are left over from when Philip Holz-mann, the big German construction company, was here. IIAA married housing are those houses along the fence over there." Andy parked the Suburban in front of a white painted concrete block building that looked a little like a Motel Six without the sign.

"Why did Holzmann leave? Must have been making a fortune here."

"They were probably kicked out by the Saudi government for not paying enough *baksheesh* to the local prince."

They got out and Mark grabbed one suitcase, Andy the other one.

"What's *baksheesh*?"

"Bribe money."

Inside, it was cool. There were large persian carpets on the floor and armchairs scattered around a lounge area. A leathery American woman of around sixty was sitting at an expensive desk with a pack of Winstons and some paperwork in front of her.

"You must be Mark," she said.

"Mark Exner," Mark said. They shook hands.

"I'm Del Winn. I'm manager of our unaccompanied personnel living facility here, the motel. I supervise our little motel here.."

"...and everybody's lives who lives in it," Andy interjected.

"Ignore him," she said to Mark. "The rest of us do." Her accent was south Texas. She stubbed one Winston out and lit another as she motioned for them to follow her out a back door into the heat and down a sidewalk between two rows of numbered doors.

Window air conditioners roared in the afternoon heat. She unlocked the door to number twelve and they went in.

"They're small but pretty comfortable." Andy set his bag down and closed the door. He removed his aviator sunglasses, revealing bloodshot blue eyes.

"Living room," Del said. "Kitchenette there." She looked over the top of her glasses. "But you'll be eating most of your meals at the mess hall like the rest of us."

"Yes, sir, boss," Andy said.

"And bedroom in here. Bathroom there." Sealy mattress on the bed, sheets and a light blanket neatly folded on it. There were towels on the rack in the bathroom.

They all sat down on the couch and chairs in the living room. The couch was a sturdy Drexel couch, obviously new.

"I stocked a few items in the refrigerator for you," Del said. "If you need anything, give me a call. We're at number four in the houses along the fence. My phone number is 2141, but most weekdays I'm up at the front desk here." She flicked ashes into an ashtray on the coffee table.

"I've got you, Danny Hager, Andy Petri, Dick Davis and Mike Robb living here full time. Plus a person or two up here from Jidda or Riyadh for a day or two at a time." She stubbed her cigarette out, stood and stuck out her hand. "We'll leave you to get settled in. Welcome aboard."

Andy shook hands too. "Chow hall opens at five. I'll come by around six and take you over for dinner

17

if you like." He paused at the door. "Tea ration isn't until the first of next month. I didn't know what you drink, but I put a bottle in the refrigerator."

He disappeared into the heat and glare.

After he closed the door, it was silent except for the whisper of the air conditioner.

Mark took his bags to the bedroom, unpacked and set the suitcases in the closet, then he stretched out on the bed and closed his eyes.

The cool clean room morphed into Allen and Carla's apartment a week ago when he'd been back in Missouri.

The four of them had been together for a long time. Mark had known Allen since high school. Mark met Jennifer as a university student, about the same time Allen met Carla. They had become best friends. They drank Busch beer, listened to oldies, read the same books, ran across each other often in the same bars and restaurants around Columbia.

Then they all graduated from the university and suddenly things were not the same any more.

"Another beer?" Allen asked Mark.

"No, I need to get back to the apartment, Jennifer's waiting."

"Remember cruising around town after class, back in high school days, in your '57 Chevy?"

"Or your brother's old green '58?" Mark countered.

"Checking out the cars and the girls."

"Been a long time, hasn't it?"

"I guess, but it also feels like just yesterday."

Mark thought of driving down quiet summer streets or dusty country roads, windows rolled down in the car, early rock on the radio, a couple of beers down, the evening air velvety, a dusky orange sun sinking behind the flat horizon. Summer evenings that never seemed to end.

"I'm having a barbecue Saturday. You guys should come over," Allen said. He downed the last of his Busch.

"Sure."

Allen ducked his head and moved his empty bottle around in the condensation on the table top. "About six o'clock. At the farm."

Mark stared at him. "The farm? You bought it?"

Allen was grinning, "Just closed on it today. That's why I was late getting here. We're moving in tomorrow."

Mark stuck out his hand and they shook hands across the table. "I'll be damned. Settling down. Becoming a property owner." He shook his head. "Next thing I know you and Carla will be getting married."

Allen shrugged, "Maybe, but we're not going to have any kids, so why add the burden of a contract to the relationship?"

"I totally agree."

The barbecue was great, as always, the talk and the jokes flowing down familiar paths. The summer evening turned to night. Jennifer and Carla went inside to walk through the house for the fourth time.

Mark and Allen continued to sit in companionable silence, beers at hand. "I've been thinking about something for a while," Mark said. "I've been with IIAA for five years, I've learned the design side of the business, now I want to get involved in the construction side."

"Aren't all the projects overseas?" Allen said.

"Yeah. I want to go overseas."

"Where?" Allen asked softly.

"Saudi Arabia," Mark said.

"Wow."

"It's a challenge I want to take before I get settled into one kind of career," Mark began explaining unnecessarily. "There are certain things that need to be done while you're a certain age. All the counter-culture stuff, the dope and the anti-war demonstrations back in the sixties it was just right for that time and place, the age we were. Now I'm ready for a change. You too. You bought this place. That's a big change."

"Jennifer looking forward to this?"

"Well, that's the thing. I just decided yesterday. I haven't told her yet."

They sat quietly, just listening to the crickets chirping in the friendly summer evening. Against the last of the glow in the western sky, Mark could see barn swallows making their graceful arcs.

"There's one thing... Allen, I want to go overseas by myself." Mark didn't meet Allen's eyes. "I think it's best for me, and her too."

It sounded false even to Mark.

Allen said nothing.

After a while, Carla and Jennifer came back out onto the deck. "You guys going to sit out here in the dark all night?"

Later before Mark and Jennifer left, Mark took Allen aside. "Don't say anything about this, okay? I haven't made up my mind yet."

Mark woke as the light beyond the drapes turned deep red-gold. He showered and put on a clean tee shirt, blue jeans, and tennis shoes. He opened his room door and the silky evening air tingled his skin.

Andy Petri was just coming up the walkway. They got in his big Suburban and drove two blocks to the mess hall. Over a meal of steak and potatoes, Mark met Mike, Tony, Frank, Dick, Danny, and Brian. The names went in one ear and out the other. Everybody was friendly.

"Don't worry about remembering everybody's names," Dick said.

"Yeah, most everybody answers to 'Hey you'," Somebody joked.

"You'll soon be sick of seeing the same old faces every day," Tony Cross said in a soft British accent.

"In a week you'll feel like you've known everybody for years," Andy said.

"And a week after that, you'll wish you'd never met any of us," Mike chimed in. He was older than the rest.

They all ate quickly. Afterwards, Andy drove him back to the motel. "I'm having some people over tonight for drinks. Like to join us?"

"No, thanks," Mark said. Andy nodded and went to his room. Mark stood in his doorway a moment enjoying the evening air.

"Hey, Mark," Danny Hager, dressed in a threadbare Kmart shirt and Wrangler jeans, leaning out of his doorway invited Mark in.

Inside Danny's cold, dingy room was a tall green bottle with a snap-over top on the coffee table. Danny pulled two beer mugs out of the freezer.

"Want you to try a little of this batch. I think you'll like it."

Mark settled himself into one of the three mismatched chairs while Danny eased the pressure seal off the bottle top. With a cough of decompression, the yeasty smell of home brew beer filled the room. With tea-ceremony care, Hager filled the mugs and recapped the bottle.

"Here's to you."

The beer had a muddy taste, and the carbonation didn't last long, but it was drinkable. Mark rubbed the ice off the Meisterbrau emblem on the mug.

"So…" Danny drawled in a central Texas accent. "How do you like it so far? Being here."

Mark smiled, tilted his chair back. "Actually, I like it pretty well."

Danny chuckled. "You don't have to be polite. Some guys hate it. One guy got here one afternoon and was on a flight out the next morning."

They talked for a while about the job.

"Andy's having a little get-together in his room tonight. Thought maybe you'd like to go. Meet some more of the people here. Them that's worth meeting, anyways."

Mark forced a smile. "Well, actually, I already told him I wasn't coming."

"Don't matter. He'll be drunk by now and won't

remember."

They downed the last of their beers, and Danny steered a reluctant Mark down the walkway to an identical room packed with people smoking, laughing and drinking.

"See that fat-ass guy?" Andy slurred into Mark's ear as he pressed a glass of gin into his hand.

The guy he was pointing at was one of the Filipino engineers IIAA hired for shop drawing review. They had been hired because they worked cheap and because they could speak and read English, not because they had any technical skills, although they doubtless had bogus but authentic-looking diplomas from technical universities in Manila.

"He's sleeping with Louis Hill's wife."

Mark pulled back to look at Petri's red eyes. "Oh, yeah?"

"Yeah. One fat-ass sleeping with another fat-ass. A match made in heaven." Andy knocked back the last of his gin and tonic. "Except that now Louis is too busy spying on those two to be able to get his work done. I end up doing it for him." Andy tried to down his drink again, realized there was nothing left but ice, and wavered off to the kitchen.

Mark drifted over toward the dining room table where Jim Redding, a Vietnamese woman, and a Brit were sitting.

"...couldn't be done," Redding was saying. They paused as Mark joined them.

"Mark, this is Ti." Jim introduced him to his wife. "I hear you and I will be working together."

"Got to get a refill," the Brit said and headed for

23

the kitchenette.

"You're on the Dormitory project?" Mark asked. "The project the other guy walked off of?"

"Yeah. I'm the project engineer."

"What happened?" Mark asked.

Jim waved his glass of Johnny Walker Red at the roomful of people. "This. Too much work, too small a community. Everybody's nose in each other's business all day and all night."

Mark drank his gin down, "This helps."

"Not always. Lots of alcoholics here. Petri for example." Redding said.

"Saudi would seem like a good place to be able to dry out with alcohol prohibited except here in the compound."

"Just the reverse. Living isolated, under pressure, no recreation, puts the focus on booze. Like this party. Everything about living in a compound out in the middle of nowhere tends to focus you inward."

Mark nodded.

"It's that way at all these overseas job sites." Redding's pale blue eyes scanned the people in the room. "I've been overseas ten years now, thirteen if you count Vietnam. Most of these guys have been overseas for years too. So long they can't go home. They're out of touch. So they just move from one overseas project to the next. Permanent expatriates." Redding stabbed his Marlboro out in a full ashtray.

After a while Mark slipped out and walked back to his room in the noiseless night air. No stars were visible in the hazy night sky.

Chapter 2

Petri picked Mark up, they had breakfast at the mess hall, then paid a courtesy visit on Bill Vance, the Tabuk area manager for IIAA. The office was sumptuous, with Persian carpets covering the floor, pale blue drapes at the windows, and potted ferns in big yellow clay Saudi pots in two corners. Ursula Cooley, Vance's secretary, rose as they entered, smiled a glacial smile, and appraised Mark with beady eyes. Andy had already advised Mark that she ran the front office with teutonic efficiency and was widely known to read all correspondence that came into the office, whether personal or business. "Welcome to Saudi Arabia," she said with a very slight German accent. She showed Mark into Vance's private office.

Bill Vance was a big man, black hair, muscles just going to fat with middle-age, boldly veined, broad chested. His expression was confident, appraising, and authoritative. "Welcome, Mark. Sorry to get you up here in such a hurry. But all our projects are right in the middle of intense mechanical systems installation and testing."

He walked Mark over to a plan view of the entire facility and went over the ten buildings under construction, "We'll place three hundred forty million dollars in four years." They stepped to a huge network schedule that covered one wall. "Here's where we are, most facilities on schedule. I won't take your time to go over everything now. You get that at the weekly staff meetings." He glanced at his watch. "I want to get you out to the Dormitory Project and let Redding get you oriented. He's a good man, one of our best. If he gives you advice, take it. But first we need to pay a courtesy call on Lieutenant Nazir, my counterpart with the Royal Saudi Military here. Protocol is important."

In the glare outside, Petri had a company Chevrolet idling. Inside was a tall young Jordanian, "This is Ali, our interpreter." Vance said. Mark shook hands with him. They drove to a gleaming new concrete and glass building near the little airfield where Mark had arrived the day before. Andy Petri and Ali waited in the car while Mark and Vance went into a large bare waiting room, comfortably air conditioned, wall to wall with persian carpet. "Lt. Nazir doesn't like having our Jordanian and Palestinian translators with us when we meet with him," Vance explained. "I believe it's official Saudi policy. They think the translators may be spies, and they are probably right. In any case, Lt. Nazir's English is perfect. He went to university in London." After a time they were shown into a very large office suite furnished in ultramodern European office furniture. A handsome young Saudi in white ghutrah and thobe shook hands with Mark.

"Welcome to Tabuk, Mr. Exner," he said in British public-school accented English. "We are pleased to have you here. Anything I can do for you, please let me know."

A young man brought a tea pot and tiny handle-less tea cups. After some courteous questions about Mark's University and hometown, Nazir spent a leisurely forty minutes discussing construction cost and schedule related topics with Bill Vance. Their tea cups were continuously refilled. The meeting seemed interminable.

Eventually they made their way to the door amid many pleasant good wishes.

"He seems like a very nice person," Mark ventured.

"He is," Vance said. "Sharp, educated, and a very tough negotiator on my quarterly payment requests. When you meet with the Saudis don't ever try to snow them. Lt. Nazir and his staff have spies everywhere. And they know a lot more about construction than they let on." Vance shrugged. "I'd do the same thing if I was in their shoes."

Mark noticed they were driving out the gate toward the town of Tabuk. "One more stop. The mayor of Tabuk."

They drove past ancient mud-block buildings and parked near a public square and went into a three story building. "His title is 'Shareef'. He's sort of a combination mayor of this town and administrator of the local area which is comprised of this town and a lot of desert. He's a distant relative of the Sauds, the royal family, like most of the civil administrators."

This time Ali came with them and they went through the waiting room routine, then were shown into an audience room. An elderly man in more formal Saudi attire rose slowly to his feet. Lengthy greetings were exchanged through the translator. Again tea was served. A slow half hour passed. Vance exhibited limitless patience. Most of the questions were simply conversational as far as Mark could tell from the translator.

Eventually they made their way out into the heat. Petri pulled the car over and they made a wide circle around the square. "Police station there," Vance pointed out. "You don't want to be in there. Remember there is no civil law here, no trial by jury, no innocent until proven guilty, just local clerics judging you on the basis of *Sharia* – the Koran-based laws."

As they pulled away, Mark saw two dark objects hanging by a strand of white cloth from a peg on the wall to one side of the police station door.

With a shock he realized it was a severed hand and foot.

"Justice swift and sure," Vance muttered. "Swift anyway."

Redding showed Mark his desk at the job site, then went over the project schedule with him. "We're about thirty percent complete. This building is a two-hundred room, four-story dormitory and classroom complex for the Saudi National Guard."

A Korean came into the job shack, respectfully removing his hard hat. Jim stubbed out his cigarette. "Mark, this is Mr. Kim, Dae Joon's general manager for this project."

The small Korean man bowed, then shook Mark's hand, "Welcome to Tabuk," he said in careful English.

Several more Koreans who had been hovering around outside came in with a pot of tea and some cups. Redding cleared project drawings off the table and the three of them sat down to tea while four Koreans stood near the door, not quite but almost standing at parade rest. Except for Mr. Kim, they all wore neatly tailored work uniforms with the company name and logo in English over the breast pocket. Mr. Kim and Jim Redding talked about project issues while Mark chafed to get to the work site and start familiarizing himself with the project. It was more than an hour before Mr. Kim and his entourage left.

"I thought you said it was a busy day today?" Mark asked Jim.

"The Koreans are very formal about certain things, and talking with the senior manager is one thing you should always make as much time for as he wants. All these guys are very familiar with IIAA procedures since the majority of the supervisors and managers worked for FED back in Korea."

"FED?"

"IIAA's Far East Division, in Korea."

"They run their companies like a military organization."

Redding got another Marlboro going. "They are, in many ways. Since the Israeli's attacked Egypt and triggered the first oil shock in 1973, the government of Korea targeted Middle East construction as their primary way to get foreign exchange, to offset the high price of oil in Korea. Almost all the construction contracts in Saudi are to Korean firms. It's not cheating; it's just that the Korean government provides financial backing and diplomatic support to Korean firms. The Korean embassy pays the performance bonds, covers the insurance, and provides de facto guarantees to the Saudi royal family that the work will be done the way they want it."

"Bribes?"

"Multi-million dollar payments to high officials are a way of life in most of the world. The Koreans know that and their diplomats bring in those briefcases full of dollars. Down in Riyadh you'll sometimes see Korean Embassy attachés going to meet with Saudi officials. Right behind them will be a couple of very tough-looking Korean men in tailored suits. It's illegal to bring weapons into Saudi." Jim held up his hand. "But those guys are well-armed. Look at the edges of their hands sometime – they can break bricks barehanded. The Korean government also guarantees there will be no trouble here among the Korean workforce. If the Saudi police complain about a Korean doing something they view as offensive, and that's a long list here, that Korean is on the next flight back to Seoul."

"Why doesn't the American government help American companies the same way?"

"The Sherman anti-trust laws. And American companies caught paying bribe money to foreign officials face indictment in the States." Jim paused. "Our government seems to bend over backward to make sure American companies can't be competitive in international construction. It's stupid and naive."

"There are a couple of American firms here in

Tabuk, right?"

Redding snorted. "Two half-assed organizations, Kendall International and McKowan. I'm damned glad I don't have to manage those contracts. Those companies send a bunch of incompetents over here, which makes us look bad in the eyes of the Saudis." He shrugged. "Well, ready to look around the job site?"

They put on their hard hats and went out into the blowing dust. Everywhere they went, the Korean workmen saluted them.

"You'll be spoiled for going back to American companies, after you've worked with the Koreans," Jim speared Mark with a look. "But don't think they won't cut corners if you're not looking; they're out to make a buck. On the other hand, when they tell you something, you can be sure it's true. They don't lie."

He leaned into the wind, then continued. "These workmen are all army privates. They get drafted into the army back in Korea, then assigned to construction companies for their army time. Now that Israel has stirred up trouble again, and the Carter administration is supporting them, we're in the midst of another oil shock. Gas shortages back in the States."

"Over a dollar a gallon." Mark said.

Redding kept on, "Like the United States, Korea imports oil from the Middle East, but unlike us they're doing fine. All these guys' monthly pay is paid back in Korea. The company receives U.S. dollars from us, they transfer it to the Bank of Korea in Seoul, the workmen are paid in Korean currency, in Korea, so it's a perfect offset for the oil Korea buys from the Saudis. I wish our government were that smart, but Carter and Secretary Vance are out promoting human rights and making it impossible for American firms to compete overseas."

"And selling F-16s to Israel at the same time."

"Giving them away, you mean. So-called off-sets," Jim grinned. "AIPAC, the Israeli lobby back home, has Congress in its pocket and most American's don't even realize it."

It was a long day, but fascinating.

The following week Jim told Mark to check the existing water line connections to some housing the IIAA had built for the Saudi National Guard. Mark drove a company Chevy over to the housing area and, after a little driving around, found the right street. Arabic numbers were easy to learn. The streets were long rows of walls with gates, behind which the houses stood.

Mark pulled the Chevy over to the wall at the side of the street, got out, and slipped his hard hat on.

The house numbers were clearly marked in Arabic, but he couldn't tell from the utility drawing whether the water line ran under the 1200 or the 1300 row.

Walled courtyards all the way down the street. He couldn't find any markings. In the States, he would have walked through backyards until he found the valve box.

He tried a courtyard gate.

Locked.

He noticed a hole in the next courtyard wall where

a truck bumper had knocked a concrete block loose. Peering through the hole, he could see some Saudi kids in thobes playing.

One boy let out a yell, pointed, scooped up a rock and threw it at him. The rest joined in.

Mark beat a hasty retreat. Even inside his car, he could hear the commotion the kids were raising. What did they think he was, an Israeli spy? The courtyard gate popped open and a knot of kids spilled out screaming and pointing. Thrown rocks peppered his car as he backed around, then drove off toward Access Road K. Dust veiled the small rage-filled faces in his rear view mirror.

Back at the project trailer, Mark described the incident to Redding, who laughed. "Welcome to Tabuk. Don't take it personally. That's just part of their culture, mostly they are very hospitable."

Mark decided to check shop drawing submittals and review Quality Control reports the rest of the morning.

After work, he showered and joined a bunch of the guys at the mess hall.

"Did you hear the news?" Frank Dray set his dinner tray beside Mark and sat down.

"No. What's that?"

"Going to be a raid tonight. Lieutenant Nazir's troops are going to search the motel."

"Oh, yeah?" Mark thoughtfully starting cutting up the lamb, mixing mashed potatoes and gravy.

"That's just a rumor," Bill said.

"Probably," Mike said.

"So what do we do with the booze?" Mark asked,

wondering where he could hide his bottle of Scotch and two bottles of Danny's home brew beer.

"Hell, I'm going to drink mine," Dick said.

"Danny, get your HSC nurses over here tonight. Let's party," somebody said.

"I don't know nobody from HSC no more," Danny drawled.

"You got any gin left?" Andy asked.

Danny laughed, eyes squinted, teeth white in his gray-yellow beard. "I don't drink that stuff."

Dick slid his chair back, stood, downed the last of his iced tea. "Your place about six?"

"Sure," Danny nodded He clapped Mark on the shoulder. "It's no big deal. We hear these rumors all the time. Nazir just wants to keep us in line. They know we drink; we know we drink. Just don't be too obvious about it. And never sell any to anybody."

Mark wolfed down the last of the canned green beans. "What time's this raid supposed to happen?"

Mike shrugged. "Who knows?"

Back at his room, Mark put his bottle of scotch in the refrigerator, dumped the home brew down the drain, then sat in his room nervously reading but no Saudi troops appeared. About ten o'clock he went to bed and slept through the night.

Mark was provided a Chevy Blazer as his company car the next morning. He drove it to the Dormitory project site and spent the day reading the plans and specs and walking the job with Jim Redding. After work, he grabbed a quick meal at the mess hall and decided to do a little exploring. He drove out the main gate and down the highway toward the Jordanian border. Sand tire tracks led off the road here and there, so on impulse, Mark turned down one, put the Blazer in four wheel drive and drove slowly back past the rocky hills through one empty valley after another. He topped the next ridge and there below him was one of the big black wool tents of the bedouin. A white Toyota pickup truck was parked nearby, ten camels tethered to one side, and several men were sitting around a small campfire. Two women dressed in black abbayahs went into the open side of the tent when they saw the Blazer. A small boy was sent up the roadway to where Mark was turning the Blazer around. The boy tapped on the glass. "Tal, ya. Shirib shai."

Mark remembered enough of his rudimentary Arabic to know the boy was inviting him to drink tea at the campsite.

He joined the men at the fire. Instead of tea they were making coffee, fresh roasted over the fire and ground in a brass hand-turned mill, it was served expertly by the boy. He held three tiny handless cups in his left hand and poured from a brass pot in his right.

Mark took the cup, "*shukran*" he said, meaning thank you. The coffee was delicious.

They sat companionably cross legged on a persian carpet that had been laid on the sand. The desert men had deeply lined faces from lives spent in the fierce desert sun. They were comfortable sitting without speaking, only the occasional crackle of the fire making any sound. The heat of the day was giving way to the silky desert evenings Mark had come to love already.

After a time Mark made it known he needed to return to Tabuk. The men stood and made courteous sounds and gestures thanking him for his company, a right hand swept across their chests. The oldest one said something to the boy who went into the tent and returned with a small silver dagger in a curved scabbard, similar to the working knives the men wore. The knife was presented to Mark. Mark exhausted his limited Arabic thanking them for the gift, then removed his belt with the IIAA logo worked into its custom brass buckle and presented it to the boy. The boy walked with him up the hill to his Blazer. Mark drove through the deepening dusk back to the road and back to the compound.

In his room he put the knife in the center of the round kitchenette table. It was obviously hand made, not perfect in any way, but beautiful in its imperfection.

The next morning was Friday, his day off. There was a pounding on his door and. Danny stuck his head in "C'mon Mark, we got to go."

"What's the rush?" Tony gestured at the half-full home brew bottles on Mark's table.

"Morning drinking out at the Holzmann camp. And some good German sausage."

"Sit down, Danny," Mark said expansively. "Enjoy the morning…good company…a selection of fine beers, custom brewed…"

"Three days ago, in a plastic waste bin…"

"…for the discriminating drinker." Danny hesitated.

"My good man," Tony stood and put his arm around Hager's shoulders. "The wise man knows 'tis better to have a bird in the hand than two birds in the bush." He steered Danny to the third chair of the dinette set, turned it out to the coffee table and seated Danny in it. "Paraphrasing Sophocles: To be happy, one must first be wise." He poured him a glass of cloudy home brew. "And this nectar of the gods will most certainly guarantee you wisdom."

"Or if not wisdom, at least a good high," Mark inserted.

"Followed immediately by a crushing hangover." Tony raised his glass. "Cheers."

They drank.

Mark refilled their glasses. "You left off the last half of the Sophocles quote."

"Oh? How's that?"

"And give the gods their due," Mark said.

"What's that supposed to mean?" Danny asked.

"That ultimately the gods give us happiness at their own capricious whim, right?" Tony didn't like not having the last word.

"Right. You can't be happy unless you're wise, and even then, that may not be sufficient," Mark said.

"Oh." Danny glanced at his watch. "I'll help you boys finish this bottle of beer, but then I'm going out to the Holzmann camp."

"You haven't even had breakfast yet," Mark reminded him, checking his watch. "The mess closes in thirty minutes."

"I ain't eating that mess hall slop again today. The Germans eat better than we do. "

"They should," Tony said. "In *der Faterland* the unions have the wage rates raised to astronomical levels. Not to mention paid-out holidays, six weeks per year, and a 35-hour work week."

"Do I detect a note of jealousy?"

"It's not sustainable," Tony's color had heightened a little. "They'll soon be where Britain is. Long lines at the dole."

"You'll all be working for us..." Danny said.

"And you'll be working for the Japanese," Tony said.

Mark raised his glass. "To universal brotherhood – all of us hating the Japanese equally."

Tony set his empty glass down and stood up. "I'm going for some food. See you guys later. Ta." He went out into the white morning.

Mark stared at the door, "Guess I must have struck a nerve, huh."

"You ought to be careful, buddy. Can't afford to piss anybody off in this fishbowl we live in. Ain't nothing to worry about. He's just pissed because the Krauts are doing so much better that the Brits. They're sensitive about that crap. I guess they never really got over World War II." Danny tipped more home brew into both their glasses.

"Hey, that's enough. I'd like to get something to eat," Mark said.

"Be plenty to eat out at the Holzmann camp. Along with German culture," Danny heaved himself to his feet. "At least German music. Funny thing, Saudi's think it all sounds the same, American and German music."

"Compared to Saudi music it does," Mark commented.

"That whining? Sounds like cats in heat."

"Music doesn't pass through culture barriers very well. Neither does humor."

They drove out of the compound to the Philip Holzmann construction company camp three kilometers away. It looked a lot like the IIAA compound, just a little older. The same desert army base feel.

The party had already begun. Danny vanished into the crowd. Mark drifted from one cluster of people to the next. Most conversations were in German; he didn't want to intrude into the English ones since he would be talking only to other Brits and Americans from his own camp.

There were four women there, three of them heavyset *hausfrau*, but there was one nice-looking woman in her twenties. She was surrounded by a crowd of guys laughing and carrying on in German.

"Who's she?" Mark asked Danny.

"Karin." He put the emphasis on the second syllable. "She's the general manager's daughter. Just out here for the summer." Danny topped off his drink from one of the bottles of clear liquid that littered tabletops. "Look, but don't touch. She's already taken."

"That's no surprise." Mark held up his glass of the clear liquid, no ice. "What is this stuff?"

"Schnapps," Danny laughed.

Ed laughed and hugged him. "Who cares!" He grabbed the elbow of a woman in an arab *jelabah*.

"Margot, my son here needs some ice. Ice for his drink. Do you have any ice?"

"Sure, we have ice." She turned pale blue eyes on Mark. "Come with me." He followed her through the labyrinthine house to the kitchen and waited until she had filled another hors d'oeuvres tray from the oven. Then she got a shallow pan of frozen water out of the freezer. "Here is ice." She handed him an ice pick.

Mark managed to chip a few flakes into his glass. Guess I offended her, he thought. He finished his drink and stood looking at the flat plain of rock and sand that stretched away to the mountains five kilometers away. After a while he dislodged Danny from the food table and they drove back to the IIAA compound.

Mark could hear Dick and Tony in Danny's room as he passed.

Early afternoon and I'm drunk, Mark thought. Think I'll lie down for a while, take a nap, then go up to the pool and swim a few laps to clear my head. He drifted into sleep in his cool blue room, and into a dream. A dream he and Jennifer had often talked about. Having their own place, deep in the Missouri countryside. He saw himself cutting firewood in the brisk autumn air, and later he and Jennifer sitting in front of the fireplace while snow swirled outside.

There was a sound, a dog howling in the distance. Mark started awake. It was dark now. The air conditioner hummed quietly, he got out of bed and turned it down. The howling came again. Sometimes feral dogs that lived out in the dumping area managed to burrow under the fence and get into camp. They were dangerous. The Saudi's shot them when they saw them. This was not a country where house pets were encouraged.

The Red Sea had some of the best reef diving in the world, Mark had been told, and it was only an hour's drive from the compound. So one afternoon Mark decided to go see for himself. He put his mask, snorkel and fins in the Blazer and headed for the beach, chuckling when he remembered his scuba instructor's emphasis on never diving alone. "Well I'm not diving, per se," Mark said to himself. He drove down the highway until he found some tire tracks leading to the coast. He saw no one and no sign of any buildings. When he reached the dunes at the edge of the water he parked the Blazer. He put on his rock shoes, slipped on his mask and snorkel and with fins in hand, waded out and slid into the clear water.

Mark drifted along the face of the reef seeing every color of red, blue, green and brown in the flickering sunlight. Behind him were only the limitless dark blue depths of the Red Sea. He rose up out of the water for a moment and looked toward the beach. The desert came right to the edge of the water, no vegetation of any kind, the flat rocky plain stopping abruptly at the waveless blue sea.

The Red Sea was surprisingly cold despite the heat of the desert all around, a rift in continental plates, narrow, deep and cold. He took a breath through his snorkel and dived down ten feet along the face of the reef. Eels with psychotic eyes stared at him from their holes.

Mark turned in the water and confronted the limitless blue. Nothing moved. It produced a deep dread in him. Below him the face of the reef sloped steeply down into darker and darker blue. He surfaced. The sun was hot. Thirty feet away, the rocky shore was comforting, his Chevy the only man-made object visible in either direction.

He again dived down the face of the reef and watched tiny orange and yellow fish hang motionless, then flit away. He worked his way back along the reef. Along the rocks at the edge of the water, the shallow pools at the top of the reef were full of scorpion fish. He stepped cautiously past them along the slippery rocks until he was in waist deep water, then jumped awkwardly into the deep water. It was exhilarating, frightening, beautiful.

He floated on his back. The sky was featureless white. If anything happened to me out here, they wouldn't find my car for months, and they'd never find my body. It was twenty kilometers from the paved road to Aqaba, the next town a hundred kilometers down the coast, and east only the empty desert. The thought passed. Mark hung motionless, breathing easily through his snorkel, letting the current move him slowly along the face of the reef. A hundred colors of fish, rock, sea urchins, and starfish passed before his face mask like a movie.

Eventually, the cold of the water drove him out. He found a rock clear of scorpion fish and hauled himself out. By the time he had clambered across the shallow pools to the sand, he was sweating in the burning sun. He put on his tee shirt, cap and sunglasses, and stood staring at the flat blue horizon for a few moments until the flies drove him back to his car and air conditioning. There was something fearful in that empty blue; a shark, a school of barracuda, anything could come out of that blue and he would be helpless. In his car, bumping along the sandy track to the road, he felt protected from the empty sky and the desert and the sea.

Chapter 3

On a bright clear Thursday morning, Mark walked to the Kendall company pool, closed the gate behind him, set his sunglasses on a deck chair, stripped off his tee shirt and sandals and dove in. The water was warm.

He swam down to the bottom and let himself drift back to the surface in the glittering silent blue. His mind also drifted, back to an earlier time.

He remembered late spring of 1969. Classes were over for the semester and most of his friends had already graduated and gone, but Mark still needed one more course to graduate, so he was enrolled in summer school and would graduate in August. Mark had convinced the draft board to extend his student deferment for three more months.

Jennifer was graduating from Christian College with an Associate of Arts the next weekend. Her parents were driving up from Florida to see her graduate, but they weren't arriving until Saturday. His roommates were both out of town, so Mark had their trailer to himself. Jennifer stayed with Mark for four days. She'd spend the night in her dorm room when her parents arrived to help her move her things out, and then she would be spending the summer in Florida before her transfer to Florida State for the fall term. But for those few days in early summer time stood still for them.

It was a breathless time. They didn't speak of the future, it was enough to just live for the moment, enjoying the hours in each other's company, sleeping late, making love, eating well. Mark read the Elizabethan poets, Gray and Marvell and Campion. Jennifer was sampling from a stack of books she'd brought from her dorm room. Mark thumbed through the stack and held up two slim books, one orange, the other blue, "Rod McKuen?"

"My roommate thought his poetry was wonderful," she said.

Mark glanced into one volume, "Sappy, but I know what he's trying to say."

Every afternoon they'd go to the swimming pool at Mark Twain dorm on campus, lie in the sun and the humid Missouri air, savor the hours in complete indolence, floating on the cool blue water under a clear blue sky. They dove into the pool to cool off and embraced against the side of the pool. "I wish these moments could last forever," he said.

"So do I," she said.

Brian and Linda Zeller clanked the gate to the Kendall pool closed, breaking Mark's reverie.

They spread their towels on lounge chairs on the other side of the pool. "Hi, Mark," Linda waved. Mark waved back and batted flies away.

"Party tonight." Brian said. He stripped off his

golf shirt, exposing a white baby fat torso. "Find an HSC nurse for your date. You'd better make your move before they're all picked over."

Linda glared at Brian, snapped her sunglasses on, and flopped down on the deck chair.

Brian grinned. "I'm just being helpful, Linda." He winked at Mark and jumped into the pool.

"We'd like to have you over for dinner one night this week," Linda said.

Mark sat on the edge of the pool. "Sure. Any night." The flies were driving him crazy.

"How about next week?"

"Sure, any time." He slid into the water, drew a deep breath, and swam to the bottom of the pool, then relaxed, letting himself drift to the surface, thinking about he and Jennifer's last weeks together, their time spent with Allen and Carla, barbecuing on the deck at Allen's farm in the long summer evening. That's what I want, Mark thought as he drifted up through blue and white. Jennifer and I dreamed about it, back in the sixties when we were undergraduates. We can still have that simple life, a real life, not clouded by money and ambition. I should have made a commitment to Jennifer, a real commitment. We might have been married now. She could have been here with me.

Mark broke the surface, and his dream again dissolved. He took another deep breath, swam to the bottom, and let himself drift up, clearing his mind of everything. He would leave Saudi Arabia when his tour was over next October, he thought, go back to the states, get back together with Jennifer. He took another breath, swam down and drifted up toward the white glare of Saudi morning. When he broke the surface Tony Cross was grinning down at him.

"Good morning," He said in a soft Welsh accent.

"Hi, Tony." Mark said noncommittally.

"I know it's your day off, but we need someone to witness the pressure test on the Training Facility water system third floor. Can you do it?"

"Ray can't get someone from his team?"

"He referred me to you," Tony said.

Mark hung on the side of the pool, two flies buzzing around his head. He let himself slip back into the clear water, then floated back to the surface. Tony was still there.

"All right. But I want Ha Li's mechanical QC man there to sign-off on the test before I do. Tell him to bring the as-builts along. We'll mark them up once we confirm we've met the requirements. Been pressurized twenty-four hours?"

Tony nodded.

"Ed Preece is responsible for quality assurance on that building," Mark said warily.

Cross grinned wider, "I hear you had a little discussion with him yesterday."

"Yeah." Mark said. "I'm not going to touch the Kendall contract. But I'll help you guys out on the Ha Li contract."

"Thanks. You're the only mechanical engineer on site," Tony said.

"I'll meet you there in thirty minutes," Mark said, then he dove down to the pool bottom again and drifted up, thinking of yesterday afternoon in Ed

Preece's office.

Redding had warned Mark not to go into the IIAA office and get into the discussion over the Kendall contract. "Let Ray Barton and Ed Preece and Danny Hager handle it, Mark. It's their contract, their responsibility. You get involved it'll come back to bite you."

"Vance asked me to talk to Ray Barton, see if I could lend a hand."

"Don't say I didn't warn you."

At the office Ray Barton had conveniently disappeared and left Mark to meet with Ed Preece.

Preece was a bearded old-timer, cantankerous and opinionated. He stayed in his office most of the time, insisting the contractors bring their issues to him. He was seldom seen on the Kendall project site.

When Mark got to his office, Ed was on the defensive and the conversation immediately deteriorated into argument.

"Look, Ed, what good does it do to continue this stand-off with Kendall?" Mark asked reasonably.

"Not a stand-off. I'm waiting for them to fulfill the terms of their contract." Ed turned and smirked at Hager, who was making a show of going over a shop drawing from the Ha Li company. "Kendall QC don't seem to be able to understand what quality control requirements are," Ed said.

"They tell me they have asked you what the deficiencies are so they can correct them."

"Who's saying that?"

"Khan and the QC guy from McKowan. I can't think of his name."

"Well, they can keep asking as far as I'm concerned. I've given them all the help I'm going to. They need to get out there and inspect the work themselves like the contract requires." His face twisted with disdain. Then he shot another smirk at Hager.

"I ain't saying nothin'. I stay out of that Kendall contract," Hager chortled. He fastidiously turned a page.

Mark tried again, "Look, Ed, all I'm asking is that we make a joint inspection. All of us together, you, me, Tony Cross and the McKowan QC man."

"They don't have no quality control men." Ed said belligerently.

"Well, shouldn't you guys see that they get one?"

"The Resident Engineer is responsible for enforcing the conditions of the contract, that's Ray Barton."

"But you and Danny are his field inspectors. Why don't we all go out to the job site and we'll go through it together. Make a list of deficiencies," Mark said.

"A punch list? That's QC's job."

"But nothing is happening. The job's been ninety-five percent done for months but the building just sits there. The Saudi's can't use it since IIAA hasn't accepted it from Kendall…"

"Kendall don't care. They've made their profit so they've pulled their people and gone home."

"So we just sit and wait? For what?"

"For Kendall to perform the work their contract requires," Ed snapped.

"And what about the Saudis waiting to get into those facilities?"

"What about them?"

"Jesus, Ed, don't we have some obligation to our customers to turn over a completed building to them on time?"

"I don't work for the Saudis. My contract is with IIAA..."

"So that entitles you to sit on your ass and wait for everything to be brought to you absolutely perfect?"

Ed's face was red. He stood up, "I got things to do, can't waste no more time with you." He turned away. Mark slapped the arms of his chair and stalked out to his Blazer, started the engine, ejected the tape, and sat staring at the Yemenese laborer running water from a hose down little channels in the sand to the row of tamarisk trees.

The office door opened and Danny Hager strolled out, adjusting his aviator sunglasses. He came over to Mark's Blazer. Mark powered the window down.

"Ed come down on you kind of hard back there," Danny drawled. "But he don't mean nothin' by it." He squinted back toward the office. "He's been burned once too many times by Kendall, trying to be helpful and all. He figures it just ain't worth it to spend time talkin' with them. They need to live up to their contract."

"So the building just sits there? That doesn't make any of us look good."

"I know." Danny grinned and slapped the Blazer lightly. "Living in the desert is hell, ain't it?" He sauntered over to his car, one of four new sedans just brought up from the office in Jidda, and drove off.

Mark rolled up the window and let the AC cool

the vehicle down while he thought. So what were his options? The tech branch wasn't going to be any help. Danny talks friendly, but he won't cross Ed. And Ray probably wouldn't back me if I tried to file a contract termination against Kendall. Kendall is looking for one more excuse to completely leave the site. Then we're left holding the bag – an unfinished building and no contractor to complete it.

A dust devil swirled closer. Its towering, twisting column was almost to the line of tamarisks the other side of the row of offices.

IIAA could accept the job as-is and try to talk the Saudis into accepting a building that's not up to par. But the Saudis could refuse, and then IIAA would be left with a building we can't complete and the customer won't accept. No options looked good.

"I'll tell Vance there's nothing I can do, unless Barton directs Kendall to do something, and that's not going to happen," Mark said out loud. The dust devil enveloped the Blazer, shook it and walked back out into the desert.

Mark drove back to his own project.

Redding was sitting with his boots on his desk, hard hat on his head. "Well?"

"I can't work with Preece."

"Told you so." Redding stubbed out his cigarette. "And now you've burned your bridges with him. He's never going to back you on any of the technical issues that come up when Vance asks for a second opinion. Preece will bad-mouth you behind your back from now on. That's why the last guy left in such a hurry."

"I learned my lesson, don't try to be helpful on the Kendall contract. I'll stay on this Dae Joon contract from now on."

"Good," Redding said. He swung his feet off the desk. "Let's go take a look around the project."

That evening, back in his room, Mark fresh from his shower, sat with his feet up on the coffee table, sipping home brew when there was a banging on his door. "Come in," Mark called.

Dick Davis opened the door and came in. "Let's go. Your date's waiting in the car."

"Not tonight Dick, I'm worn out."

"Sorry, no excuses." Tony Cross followed Dick in the door and they hoisted Mark to his feet.

Dick drank the rest of the beer in Mark's glass while Mark put his shoes on.

In Dick's Suburban were three nurses from the hospital. They introduced themselves on the ride to the Kendall rec center. The last one to lean over the seat and shake hands with him was a strawberry blonde with blue eyes, perfect teeth and a quick smile that disappeared just as quickly. Her handshake was strong and the skin of her hands a little rough. "I'm Cathy Locke," she said. "From New Hampshire." The other two were Elspeth and Moira, from Ireland. As they chatted, Mark was a little put-off by Cathy's quick gestures, almost jerky, and her tendency to make over-positive statements. She had a habit of shaking her head in three quick nods. But she was pleasant, they all were. At the rec center they stepped out of Dick's Suburban into a soft evening. The white light bulbs around the patio fence had been changed to blue. Along with the pool's underwater lights, they provided a cool blue glow to the dozen people sitting and standing round the tables.

Somebody cut off the cassette player halfway through Del Shannon's *Little Town Flirt* and started some Glenn Miller 40's dance music. Dick Davis pulled one of the nurses into an awkward version of West Coast swing, dangerously close to the edge of the pool. Tony and one of the Irish girls were doing some sort of current UK dance that looked even less graceful. "Would you like to dance?" Mark asked Cathy.

"No," she said. "I don't dance, but thanks."

"Then let me buy you a free drink." He led her over to the makeshift bar and poured some home brew beer into two glasses. "Skoal."

Brian staggered over, "Hello, hello." He tried to hug Cathy, who dodged him, except for a kind of shoulder clutch.

"This is Brian Zeller," Mark said. "It's his birthday. His party."

"Congratulations," she said.

"Think nothing of it," Brian slurred, "I don't." His wife Linda sat in one of the lounge chairs, watching coldly. "I'd introduce you to the rest of the people here," Brian continued. "But it's not worth it. None of them are any good."

Cathy laughed uncertainly.

"Just you and me, right?" Mark joked at Brian.

"Some people have their doubts about you," Brian said.

Cathy looked puzzled.

"Preece, for instance," Brian slurred. It took him two tries to pronounce 'instance'.

"Is he here tonight?"

"No. On leave." Brian was staring at Cathy's breasts.

Mark pulled Cathy away. "Let's go see who else is here. We'll see you later, Brian." They made their way around the pool. "Sorry about that. He's drunk. He's often drunk. Unhappy in his marriage."

"Obviously." They came up to a tight cluster of people at a corner of the pool. "This is Tony Cross, Kendall International's quality control chief."

Tony stopped his story and shook hands with Cathy. "Hello, hello, love," he said affably.

"And this is Jim Redding and his wife Ti," Mark continued.

Jim's only concession to the festive occasion had been to exchange his work boots for a pair of clean white sneakers. The worn cowboy shirt and jeans were the same. Surprised he didn't wear his hard hat tonight, Mark thought.

"Continue, Tony," Jim said.

Tony looked embarrassed, "Well…"

Cathy laughed. "I've heard off-color jokes before, believe me, so go ahead. I don't embarrass easily."

"Right. Well, as I was saying, the one lad says to the other, 'lovely face but no bristols' and the whole pub fell down laughing."

"Bristols?"

Tony held his hands in front of his chest, "Breasts, women's breasts. There was a dancer in Britain, Doris Davenport, fabulous body. The old gutter rats,

54

dirty old men, started calling her breasts 'bristols'. Don't know why. Just another 'leering cross the park at the young girls' kind of phrase, but it stuck."

Tony grinned. "So now one of the largest air compressor firms in Britain is running double page ads in the tabloids, Doris posed between two big compressors with a slogan reading 'the best bristols in Britain.' Huge success. Sales have shot up."

A very pale boy was tugging at Tony's sleeve. His face was a an exact, younger, version of Tony's.

"This is my son, Albert."

"Mother says the water has gone off again and can you fix it."

Tony laughed good-naturedly, "Duty calls. Ta." He and his son made their way around the pool. The boy walked with exactly the same gait as his father.

Ray Barton sauntered up. "Hello Mark. Can you come over to the Training Facility and check some fire control piping?"

"What's up?" Mark asked cautiously. Ray was senior to Mark, he'd try to do what he asked, but not if it involved the Kendall contract.

"We need the Halon fire control system checked ASAP, and nobody's qualified except you. You got your Professional Engineer license yet?"

"Not yet. I've passed the Engineer-in-Training exam. I need five year's certified experience."

"Well, you'll get plenty of experience around here," Dick Davis had come up and was topping off everyone's glass.

"When do you want to run the tests?" Mark asked Barton.

Barton gave Mark a guileless look. "Tomorrow?"

"Friday? My one day off?"

Barton nodded. "It's on the critical path. The Koreans had air freighted in the parts, and the tech rep from New York has checked it out. He'll stay tomorrow if you'll come out and look at it. Otherwise he's going back to New York."

"Alright," Mark said. "Have the tech rep and the Ha Li guys meet me at the field office at around nine a.m."

Barton shook Mark's hand "Thanks, I'll..." He was interrupted by raised voices at the other end of the pool.

"What the hell for?" Brian shouted at Linda. They stood toe to toe.

Dick staggered over to break up the argument.

Mark slid Cathy behind him. Elspeth and Moira came over.

"You have no right to embarrass me!" Linda shouted at Brian. "Always after these women."

"I didn't do nothing," Brian whined. Dick laid a hand on Brian's shoulder, trying to calm him. But Brian twisted away. "Say out of this."

Linda pointed an accusing finger at Brian. "last week, and now here in front of everybody. I've had it with you. I'm leaving!" she stomped out the gate banging it shut behind her.

Brian didn't go after her. Dick hovered over Brian apparently trying to calm him. But Brian shoved him away. "Get the hell away from me. I think it's pretty chickenshit to go pushing people around. I should never have invited you."

Dick reached out to conciliate but Brian swept his arm up, knocking his own glasses off. Dick went off balance and fell into one of the white plastic tables which went over, spilling glassware. Brian's glasses skittered to the edge of the pool and into the water.

"Son of a bitch!" Brian shouted at Dick "Don't shove people around." Brian staggered to the edge of the pool, looking for his glasses. Floyd, who'd been sitting on a chaise lounge drinking wine all night staggered up and announced, "I'll get your glasses for you." He knelt by the pool, reached into the water and slowly rotated into the pool. He came up sputtering. Everyone around the pool was roaring with laughter. Brian stood there, looking around belligerently, then helped himself to another drink before he sat down heavily in a deck chair.

Floyd stood up in the shallow water. A weak smile spread across his face. He sank slowly under the water and held up the missing glasses.

"I guess the entertainment's over," Mark said to the three women. "Dick's in no condition to drive. Can I drive you back to the HSC dorm?"

The four of them left together and walked back to the motel where Mark's Blazer was parked. Cathy took his arm as they walked. Elspeth and Moira walked behind. "Sorry about that," Mark said.

Elspeth laughed, "You should see the odd lot we have at the hospital."

"I'd say Brian and Linda are long overdue for a divorce," Cathy said. "And living in these tight little communities will finish it off if it's not stable to start with."

"Likely they stayed married out of inertia," Mark added. "Brian's like that on the job. Afraid to make a change, so he never makes a decision. That's one thing I like about Redding, the guy I work for. He's right in the middle of all the contract decisions. Real decisive. The Koreans respect him even when they don't agree with him." They got in Mark's Blazer and drove to the HSC compound. Mark eased the Blazer over the last set of speed bumps at the HSC gate. The bedouin gate guard flicked his wrist, a throw-away gesture, meaning they could pass.

"Thanks for the ride," all three women said at once and piled out.

Mark spent all the next day working through the fire control system at the Training Building with the CombustionTech representative from New York. The next three days were spent inspecting air conditioning rough-ins on the Dormitory building. The days flew by. At lunch at the mess hall Thursday, Dick grinned at him, "You're working too hard, little buddy."

"I need to get prepped for the water system checks next week. I'm going to work this afternoon," Mark said. It was a big and elaborate system.

"No you don't. You need to come with me, over to Aqaba. It's the Red Sea resort town just across the border in Jordan. Booze is legal there," Dick grinned.

Mark pushed his empty plate away and leaned back in his chair. He was tired. "When can we get back?'

"Friday afternoon," Dick said. "It's about a four hour drive each way."

"Well...sure," Mark said slowly. He didn't care about the liquor but he needed to get away from the job site for a day at least, to clear his head. "When do we leave?"

"Right now," Dick said with his big smile. "go pack an overnight bag and get your passport. I'll meet you in front of your room."

They drove through empty desert, then past sun-blasted mountains. The highway was perfect, new and traffic-free except for the occasional pick-up truck going either eighty miles per hour or twenty miles per hour. They came through the last pass through the mountains and there was the deep blue Gulf of Aqaba. They checked in to the Holiday Inn. Dick dragged Mark next door to the Aquamarina Hotel's upstairs bar, where he ordered a double Johnny Walker Black, drank it down, and ordered another one. He looked at the gold Patek Phillipe watch on his wrist. "Got lots of time. Bar doesn't close until midnight."

He knocked back his second drink and signaled for another. The Jordanian bartender gave him the eye but poured him a Johnny Walker black, neat. "When we get back you should look up that little blonde I introduced you to the other night. The nurse from HSC." He smiled his creamy smile.

"Nice looking girl," Mark said.

"I used to date a blonde," Dick said. "Long time ago. Back in Seattle." His chuckle turned into a cough. He tapped a Marlboro out of his pack and lit it. "Had it made there for a while."

Mark ordered an Amstel and settled in, "And

then?"

"Decided to sell my company and go overseas. Work for Intercontinent in Korea. Lots of companies there doing construction work for the U.S. Army in Korea." He squinted at the mirror back of the bar. The gold and red lights made his puffy face look healthy. "Thing about Korea is, it's a man's society. American women don't fit it in, and the temptations for American guys are..."

"I didn't know you were married," Mark said.

Dick grinned, "Went to Korea married, came back single. Lots of guys did. I stayed six years, 1958 to 1964."

"You had your own company before that? That's great," Mark said. "I'd like to have my own company some day."

"Careful what you wish for."

"General contractor?"

"No, an injection grouting firm." Dick waved his cigarette around. "Having your own company is too much work. I'd rather work for Frank Dray, like I do now. Frank's company, Intercontinent, is a contractor to IIAA, supplying engineers to IIAA. I've been with Frank for almost eight years. He's got a little one-desk office in an office park out in Hawaii Kai. Multi-million dollar firm, hundred people on the payroll but they're all a bunch of engineers in Manila, five hundred dollars a month for those guys. Us field guys only work for a year or eighteen months. I work until I've got fifty thousand in the bank, then I tell Frank I need some time off. I go back to Honolulu, relax until the money's gone, then back to work.

A bunch of us been doing it for years. Between jobs we all hang out at the Down Under, a little bar on the lower level of the Ala Moana shopping center, the old wing. I've got a condo in Waikiki, no car, walk everywhere. Get up at noon, go down to the hotel bars along the beach and talk to the tourist ladies."

"Kind of hard to stay in touch if you're gone for a year at a time."

Dick laughed. "Not really. All us guys, construction guys, we see each other here and there, all over the world. It's a big family. Spend a year in Honolulu, go back to Tehran on some job and it's like I never left. Construction camps are all pretty much the same. So are the jobs." Dick ran his finger down the polished wood of the bar, a complex emotion on his face.

Mark finished his Amstel. "Sounds like you've still got it made."

Ed looked at the amber liquid in his glass. "Yeah. For a kid from an orphanage in Spokane. No mother no father, just me. I worked, finished high school, went to work in construction, worked hard. I knew what I wanted. I wanted a convertible and a blonde. Got them both, too. Happiest days of my life, driving around town in my Buick with the top down, Mary Ann – that was her name – riding beside me. I saved my money, borrowed some more and got my own little company started. Not even thirty years old and I had it all. Mary Ann..." Dick smiled at something only he could see.

"Well, I'm going to wander around a little. Explore Aqaba," Mark said.

61

"That won't take long," Dick finished his drink. "Guess I'll go down to the Sheraton bar, see what's happening there." Mark noticed Dick had tucked a twenty Jordanian dinar note, worth about thirty dollars, under his coaster.

"That's a lot of money for a tip," Mark said.

"Treat people right. I've learned that," Dick said with a brilliant smile. "It's only money."

The next day Mark woke up feeling better than he had in a long time. He lay in the comfortable king sized bed listening to the gentle hum of the air-conditioner. Through the sheers he could see the inviting blue pool and the row of blue deck chairs, sparkling clean looking in the white light of morning. He slipped on his swimming suit and went out to a deck chair in the sun and admired the cobalt blue of the gulf and the rugged mountains, faintly lavender and dusty yellow, rising up to the white sky above the beach town of Eilat. After a while Dick came out, hung over, tanned, overweight, laden with gold chains. Another guy about the same age as Dick wandered over from a room at the end of the row. "I ran across Bill Riley last night. We used to work together in Guam." They laid out their towels, their cigarettes, their lighters and ordered bloody marys. After their second round of drinks, they all ordered breakfast. Mark was hungry enough to relish the runny eggs and imitation sausages.

Another round of drinks was ordered – Dick and Bill were reminiscing about past construction jobs, in Vietnam, in Venezuela, in Germany. The alcohol rose to Mark's head. The stories were fascinating.

They had been everywhere and done everything. I want to do all that, Mark said to himself. See the world, make money and spend it, work and live and party in the most exotic places.

"But you know what the best thing is, the really best thing," Bill slurred.

"The women," Dick said.

"Well, yeah...no." Bill shook his head, took a long drink from his glass which was empty. He signalled the waiter. "No, it's the friends you make in this business. That's the best thing. All over the world, we all know each other, we've all worked together here and there, go into any airport bar and you'll find somebody you know, or somebody who knows somebody you know. It's a fraternity. It's family."

Fresh drinks arrived and they raised them in a toast. "To friends."

Dick stared at his glass of scotch "Including absent friends." They raised their glasses again.

The breeze was refreshingly cool coming up from the cold Gulf.

"Well, I'm going to go rent a jet ski," Mark announced.

"Hey Mark," Dick grinned under his sunglasses. "Elspeth over at the hospital said her friend Cathy likes you. You need to go see her." Dick and Bill laughed.

"Oh yeah?" Mark said. "I will. Thanks." His head was spinning form the morning drinking, but he managed to get down the stairs to the beach rental stand. He gave the guy five dinars, told him he knew how to ride, and got onto a purple and red Kawasaki jet ski.

He idled out into deeper water, then twisted the throttle wide open and raced toward the row of freighters a mile out. Behind him a twenty foot rooster tail of white water arced into the brilliant sun.

He slalomed around the slow-moving barges bringing in tank after tank, long barreled T-72 tanks manufactured in Romania. At the dock, there was the constant rumble of diesel engines as the tanks were started up, driven onto flatbeds pulled by an endless line of trucks, and hauled up the highway to Syria.

Back in Tabuk, Mark worked fourteen hour days for the next four days. Tuesday after work he hurriedly showered and drove over to Floyd and Danielle's house for dinner. He hadn't really wanted to go, but Danielle had invited him, so here he was. He stood at the front door with his hands jammed nervously in his jeans pockets. Finally he rang the doorbell and stepped back a step.

"Hey! Come on in, come on in." Floyd shook his hand and pulled him into the house. Delicious food smells filled the house.

Floyd's jittery friendliness contrasted to his taciturn French wife, just as his round-headed smiling face contrasted with her hawk-nosed impassiveness. Mark learned later she was a war bride from impoverished 1945 France.

"This is Andre," Floyd introduced a sun-darkened European sitting comfortably in the big chair drinking wine with a sophisticated air. He stood and shook hands with Mark.

Mark accepted a glass of home brew wine in a

German glass from Floyd.

"Andre's with a Belgian firm building facilities for the Saudi tank corps out in a compound out toward Tabuk." Floyd jumped up, poured a microscopic bit more wine in everyone's glass. "Andre's our competition."

"Not competition so much as predecessor. We're on our way out. You Americans are the current favorites," Andre said with a smile.

"The Saudis never like to tie themselves too tightly to any one nation. Before the French it was the Germans," Floyd added.

"And after the Americans will come the Koreans," Andre said with a laugh.

They talked shop until dinner was served. After an excellent beef and salad dinner, they adjourned to the living room. Floyd refilled all their glasses.

"It's better," Andre said.

"The wine?"

"No. The equipment. The training equipment we are installing in Building 20 is better than what the French army itself has."

Mark nodded. "Well, the Saudis have more money. I see oil is $37 a barrel."

"So they can afford all these new toys. Like the electronic simulator in Building 20."

"And all the M60A3 tanks they bought from the United States..."

"There's a bunch of British Chieftain tanks in that row of steel buildings along Perimeter Road," Floyd added. "Must be a couple of hundred."

"All those sheet metal buildings on the road toward the water treatment plant are full of tanks," Andre said. "I was out there one day after they had delivered a bunch of them. A couple of Saudi troops were parking them." A smile crinkled his face. "Looked like fun; hot rodding those big, noisy new toys around empty lots. I'd like to drive one some-day."

Andre took a packet of Craven A's out of the pocket of his military-cut shirt and lit one with a gold Dunhill lighter. The window air conditioner rumbled on. Danielle joined them.

"Think these guys could ever actually win a war?" Floyd asked Andre.

Andre snorted, smiled, shrugged.

Mark held his glass up, giving it an appreciative look. "This stuff is starting to taste a little better," he said, then returned to the previous conversation. "I don't know much about military strategy, but there are so few trained Saudi troops it seems unlikely they could win a war against, say Iraq or Syria or Iran. They just don't have that many people in their army."

"Or in their whole country," Andre added.

"How many Saudis are there?" Floyd asked.

"Nobody knows. And I doubt the Saudis will conduct a census. They value their privacy too much, and they worry their enemies would somehow use the information against them."

"I've heard estimates of from six million to fifteen million Saudis," Floyd said.

"And an equal number of foreign workers."

"Foreign workers are necessary for the moment.

But many of them want to stay." Andre said. "The Saudis require them to leave when their work is done. That's one purpose of the exit visa we all have. The Saudis don't have an army yet, and they're surrounded by enemies. But they don't want to risk hiring too many foreign soldiers..."

"Because foreign troops could take over the country. It's happened often enough in the past." Floyd finished for him.

Danielle lapsed into French for a moment with Andre.

"By the way, Mark," Floyd said. "Danielle and I are going to Damascus and Istanbul next month. Would you like to go along?"

"Sure," Mark said without giving it much thought.

It was a pleasant evening. Mark was home and in bed early.

The next morning Jim Redding stomped into the job shack, clearly not in the best of moods.

"Trouble?" Mark asked.

"Vance dumped this Kendall change order in my lap." He dropped a fat folder on his desk, sat down and put his work boots up on it. "Our incompetent clods and Kendall's incompetent clods have been changing the HSC rec center mechanical systems without updating the as-builts, so now nobody knows what's out there, much less who owes who how much for all the changes from the original design."

Mark slid the folder out and took it to his desk. "Let me take a look at it."

"Now Vance wants the change orders settled this

week, and of course Hager, Cooley, and your buddy, Preece are going to think of every excuse not to. That's why Vance gave it to me, to see if I could go around them and get Kendall to settle. I'm supposed to meet with their office admin guy, Kurt Hess, tomorrow."

Mark spent the day working through the drawings, went over to the rec center and sketched how the mechanical room looked now, then highlighted a copy of the mechanical room contract drawing to show the changes that had been made.

Redding said nothing.

Late afternoon, Mark made a hurried walkthrough of the second floor duct work fabrication with Mr. Pak, the mechanical quality control supervisor. They worked out several issues and Mark annotated the as-built drawings for that section, then went back to his desk and filled out his quality assurance report.

He grabbed a quick sandwich at the mess hall, then went back to the job shack and did a take-off and estimate of the mechanical system changes on the rec center. It was ten at night by the time he laid the competed estimate and the marked-up drawings on Redding's desk.

The next morning Redding was already in the job shack when Mark got there.

"This is good," he said, "Thanks. I was afraid I'd be going in to negotiate with the German ogre empty-handed." He thumbed through the papers. "Bottom line is seventy-five thousand dollars and ten days. No way to tell who's responsible for these changes, them or us?"

"Somebody would have to go through all the past correspondence and QA/QC reports. Cooley and Preece should have the correspondence file."

Redding snorted, "Not likely."

Mark grinned, "Well, I think I've identified what changes have been made at least. Maybe you and the German can reach agreement on what has been done, and what it cost, then let the area office people work out who's going to pay." He nodded at the estimate, "That's the best I could do with no manufacturer costs available. Some of it is guesswork."

"Better than nothing." Redding glanced at his watch, "Well, gotta go." He went to the door and opened it, admitting a gust of oven-hot air hazy with dust. "Why don't you come along? It will be good experience."

"Sounds interesting," Mark said. He got up and followed Redding out to the company vehicle.

The Kendall office was a white sheet metal building shaded by tamarisks. A big Kendall International logo had been painted on the wall. Inside it was luxurious expensive persian carpets, ferns in brass pots, silent air conditioning and double pane windows.

Kurt Hess was standing in the middle of a large persian rug in front of the reception desk.

He was a medium-sized man, in top physical condition, perhaps sixty-five years old, bald on top with a fringe of white hair. He was dressed in a khaki uniform bare of any ornamentation, but Dick Davis had told Mark that Hess had been in the German army in World War II and a uniform seemed natural on him. Blue-eyed and big-featured, Kurt grinned, showing square white teeth while giving Jim and Mark a bone-crushing handshake.

"Come into my office." He motioned them into a large office, bare to the point of starkness. There was a slim folder lying in the middle of his otherwise clean desk.

"We are here to discuss the changes the IIAA has directed in contract 0236," he began.

"We are here to discuss the mechanical systems at the HSC rec center," Redding interrupted mildly.

Hess gave them a glittering blue-eyed glare. "I will start at the bottom line. IIAA owes Kendall International seven hundred fifty six thousand dollars and eighty-nine days time on contract 0236 for all directed changes to date."

"We are here to discuss the rec center mechanical systems only," Redding repeated.

Hess opened his folder and pulled out two drawings and a quarter-inch thick sheaf of official correspondence. Mark saw IIAA letterhead on most of the sheets.

"The bottom line for the rec center mechanical changes is one hundred fifteen thousand dollars and twenty-three days," Hess said flatly.

The conversation went on for almost another hour, Redding trying one approach after another to get started on a compromise, but nothing worked.

Hess had all the facts, who had directed what, when they had done it, how much impact it had on their schedule, what the cost of materials were, what the crew size and skill mix doing the work, and what material and equipment shipping charges were. All the field changes derived from direction given by Cooley, Zeller, Hager, or Barton.

Redding continued calmly, trying tack after tack, but Mark could see the back of his neck getting red.

"Well," Redding said finally, we came here to see if we could reach an agreement and apparently we cannot."

Mark followed him out.

In Redding's car on the drive back to the job shack, Mark shrugged. "Sorry about that. I didn't give you much to work with."

"Not your fault." Redding nodded in the direction of the IIAA field office as they drove past it. "Those guys in there have got that contract so screwed up, all we can do now is offer a global settlement to fix it."

"Global?"

"IIAA and Kendall each comes up with a number and negotiate until they reach an agreement, regardless of the history of the changes. I'm sure that's what Kendall is waiting for. Our folks too, since it will be easier on them. They won't have to do their homework of sorting out the history of this mess and preparing a solid government position. Just pay the contractor what he asks and get rid of him. It's the easy way out."

"That guy Hess is one tough son of a bitch," Mark said thoughtfully.

Redding nodded, "That's his job. We argue during the day, but after hours we're all friends. Nobody can stay isolated in this tiny compound."

"Sure is a lot easier dealing with the Koreans."

"They'll spoil you," Redding said.

Chapter 4

"Well, it ain't Lone Star beer, but it does the job," Danny Hager said, topping up people's glasses with murky home brew.

It was the usual after-work gathering in Danny's dingy room: Dick and Tony, Mike, Ray, Brian. Even Floyd had dropped by for a change. This stuff tastes pretty good, Mark thought, that's a bad sign.

"I was home last year on vacation," Danny shouted from the kitchen. "I talked to the guys I used to work with back there. I'm livin' better than they are: free car, free housing, cheap meals. I can save a ton of money. They got houses and cars and big TVs, but ain't none of it paid for."

Floyd was jittering around the room, straightening up the mess Danny lived in. "There are a lot of good features to compound living overseas. Saving money is one of them. Also, people get to know you, they help you, mostly. Not like back in the States. When we moved to Jacksonville, we didn't meet our next door neighbors for six months. Never did get to know them, not really."

Dick tilted his glass of scotch, "Around here you get to know your neighbors real well, real fast. Maybe too well." He winked at Mark. Danny collected home brew bottles and took them to the sink.

"Ten down, one hundred twenty to go."

He brought out another bottle and started to ease the retainer ring off the bottle top. It snapped open, and foam sprayed over the coffee table. "It's hard to control the pressure in these bottles. Lots of times they'll blow up just sitting in a box. Glass slivers everywhere."

Someone knocked on the front door once, then burst in. Mark started and made a half-hearted effort to hide his beer. Just like college days, he thought, hiding our beers when someone comes in. A young guy dressed in U.S. Army fatigues with a military haircut stood in the doorway letting the heat in. "Danny! Let me borrow your car. My goddam battery is dead again."

Danny tossed a thick ring of keys to him, "How 'bout a beer before you go?" Danny said mildly.

"That dog piss you brew?"

"Come on now."

"No thanks, I've got to get cleaned up and go pick up Moira. See you guys." The army guy closed the door on the gold evening light, then ducked back in. "What time you need the car back?"

"It don't matter."

The door closed. Hager tilted back in his chair and began to ease open the snap ring on a fresh bottle of beer. An ominous hiss issued from the bottle.

"Where'd you get that second refrigerator?" Mark asked.

Danny smirked. "From the executive assistant who worked here before Andy Petri."

"I need to talk to Andy, or Del Winn," Mark said. "I'd like to have one myself."

"Don't bother, they ain't shit," Danny snapped, pouring everybody's glass full again. "They'd like to think they are, but they ain't. They were nobodies back in the States. Then they come over here, get promoted, and think they're running the place." Danny slugged down a generous gulp of beer. "I been here four years, four years, man. I seen their kind come and go. Some good some bad. These two are worthless."

Mark's stomach was growling. He drank his beer down to the sludge line and made his way through strewn clothing and the tangle of mismatched furniture to the door. On the back of the front door, Danny had taped up a Lone Star beer poster, the only decoration in the room. "I'm going over to the mess hall for dinner."

"How about giving me a ride over there," Hager said. "The rest of you guys stay here as long as you want, make yourself at home."

"We will."

They stepped out onto the warm sidewalk under a vault of royal blue sky. The air had the silky-smooth feel of desert evening. Mark unlocked his Blazer, climbed in, started it up and cloud of dust came out of the air conditioning vents. Once the air cleared and Mark stopped sneezing, Danny heaved himself into the Blazer.

"Still locking your car?" Hager asked laconically as they drove.

"Just habit, I guess."

"I don't even lock my room any more, much less my car. Ain't no need, no theft here."

Mark headed for the Kendall mess hall.

"No, turn here," Danny said. "It's time you saw a little bit more of our compound. Let's eat over at the HSC cafeteria."

He grubbed around in his blue jeans pockets and pulled out a wad of five and ten riyal notes. "Just wanted to make sure I had some money with me. Not many expenses out here, so I hardly carry any money with me."

Mark parked in front of a white wall with a pedestrian gate beside the HSC logo. Inside the compound, the recreation center was to the left, the hospital in the center, the dorm to the right. Beyond the roof of the two-story dorm, the distant mountains rose purple and gold in the dying light. The moon was a shining white disk in the purple sky. Danny gave Mark a shove forward to stop his gawking. Inside the glass double door of the recreation center there were a dozen men at several tables; Jordanians and Palestinians mostly, a few Indonesians, Filipinos, and Egyptians, the usual mix. All the men in the place were glancing from time to time at the table near the back wall where two young western women in hospital uniforms were eating and talking.

Mark got a tray and followed Danny through the serving line.

"We built this building two years ago. It was one of my projects." Danny signaled the Palestinian kid behind the serving line for more rice. "Kendall had the contract. This building come out pretty good, didn't it?"

Mark glanced back at the bright yellow-and-white painted concrete block walls in the dining area, the red plastic tables and chairs, the stainless steel serving line, the quarry tile floor. "Looks good."

"Mostly American stuff too, the kitchen equipment and all. Contractor had to pay a lot of baksheesh, 'expediting' money, to get things through customs on time. Just a cost of doing business here, I guess."

Mark went down the cafeteria line, got rice and stew, some kind of mixed vegetables and a Pepsi. It came to eight riyals, about two dollars.

Mark stuffed his change into his jeans pocket and followed Danny over to the table beside the women. They turned out to be Cathy Locke and Mary, a colleague of hers. Mary was one of the older nurses, deeply tanned and physically fit. Mark learned later that she had been a mentor to Cathy during her time at the HSC hospital.

"How ya'll doing this evening?" Hager drawled.

"Just fine, thank you." Mary said crisply.

"So much to do in the evenings," Cathy laughed. She had pretty blue eyes, Mark noticed. Her face was pretty – not beautiful, but pretty.

"Pool looks nice," Mark said, gesturing at the big blue swimming pool outside the glass wall. "You could go swimming." The evening sky above the wall was a deep royal blue; the underwater lights glowed emerald blue. It looked almost unreal it was so perfect.

The two women looked at each other. "It's not women's swimming hours now," Cathy said.

"Swimming hours for us are four to six," Mary added. "That's when they close the building and run all the men out of here so women can swim."

"Like we'd be attracted to that lot," Cathy added derisively, tilting her head at the arab men sitting, staring.

"That reminds me," Hager said to Mark. "Kendall's been complaining we've got a design bust on the pool pumps. Number two pump keeps tripping circuit breakers. Their folks have to come over here to make service calls about every week. Maintenance is not in their contract. Hows about you check it out."

"That's Preece's problem," Mark said shortly. "But if you'll get me the submittals and the mechanical shop drawings, I'll take a look."

The women excused themselves, "We're going on shift now."

Mark and Danny finished eating. "Cathy's a nice looking girl," Mark said.

Danny leered, "Interested?"

Mark shrugged, "Maybe."

The next day, after checking the Dormitory building job site and project schedule and discussing progress with Mr. Kim, Mark drove over to the IIAA office, Hager had his boots up on the desk, as usual. Preece was going over some submittals.

"Hi Danny," Mark said, ignoring Preece. "Got those pool pump shop drawings?"

Hager waved at a row of file cabinets. While

Preece pretended not to watch, Mark went to the files, found the mechanical systems section and eventually found the pool piping diagrams and manufacturer's literature on the pumps. Preece's signature was in the approval block.

"I'll bring these back in a couple of days."

"Sign-out sheet's over there," Hager drawled.

Mark signed for them and drove around to the back of the HSC rec center, where the mechanical room doors were. There was a white Chevy Suburban with Kendall International logo parked there. Three McKowan mechanical subcontractors, none wearing hard hats, were in the mechanical room.

"That's got it. Bolt her up," one said. The floor was a litter of hand tools.

"You guys working the pump problem?" Mark asked.

"Fixing your design problem, you mean," one of them snickered.

"On the job, keep your hard hats on," Mark said.

They ignored him. One flipped the breaker back on and the pump started up with a whine. "Don't need them. Nothing's going to fall on us here." They gathered up their tools. "Let's leave our IIAA man to stand here watching the motor run." His buddies grinned. They sauntered out and drove away.

Bastards, Mark thought. Kendall and McKowan, their subcontractor, have already made all the profit they are going to make on this project. They've moved their best people on to other jobs and left these incompetents to fix the last remaining problems. Compared to the Koreans, these American workers are worthless. Companies like Kendall and McKowan make their profits in claims court, not on the job like honest companies should. The Saudis see these clods and it makes all Americans look bad. When the Saudis contract directly with a company, if the contractor doesn't perform, they terminate the contract and get somebody else, quick and clean.

Mark got to work, comparing the constructed piping to the approved shop drawing. Everything looked all right. He put his ear to the pipe; there was the faint whine of cavitation starting again. He pulled out the pump manufacturer's manual and went through the installation requirements – base plate mass, vibration isolation, flex connections, wiring the right size and connected right. He went to the panel. Breakers were right. The filters had a transparent sediment bowl. It was clear. The sand filter bed looked clear enough.

The cavitation whine deepened. Mark went out the front door of the mechanical room onto the sunny pool deck. A Yemeni was skimming the pool and stopped to watch. Where the water ran back over the gutters to the pump inlet there was a metal cover over the pump well. Mark slid it back. There was a whirlpool forming in front of the number two pump inlet. The cavitation whine increased. The whirlpool broke the water surface and the pump sucked air, chugged for a moment trying to pump air, then shut down on overload. The number one pump continued running smoothly.

Mark went back into the mechanical room and stared at the piping again. The pumps were offset from each other in their parallel piping, for ease of access. Mark had a sudden thought and thumbed back through the manufacturer's literature, then measured the straight line length in front of the number two pump. That's it. Less than four diameters straight run of pipe in front of the inlet, not enough length to let the water get into laminar flow before it enters the pump. The water stays turbulent, the impellor cavitates, sucks in a whirlpool, stalls and shuts down. Mark put his tape measure on the inlet pipe. Looks solvable, unbolt the pump flange and weld in a longer piece of line – there's room enough – then up the wall to the inlet at the pool gutter. Mark sketched the solution and wrote in the dimensions.

Mark got back in his Blazer and drove to his own job site and spent the day working out piping issues on the Dormitory water supply. He didn't leave the job site until almost eight o'clock that night.

Next morning he was back at the Dorm project by seven o'clock, his blue golf shirt already wet with sweat at his lower back. The air temperature was just rising past ninety. He unrolled the shop drawings he brought with him on the concrete floor. With tape measure and flashlight, he identified the 3/4" copper water supply lines, checked the stripe color for type: red for hardened. He had already found an empty solder spool in the last room so knew they were using 90/10, not 50/50, solder. Or maybe they'd just seeded the area with empty spools. He wouldn't put it past them.

He slid between the open steel studs and laid his tape from the piping floor penetration to where it was stubbed-up where the lavatory would be. Okay, a meter four. Cold on the right. He traced the toilet supply from T off the main to the stub through the wall. He pulled a copy of American Standard's rough-in drawing out of his hip pocket, checked the stub-in height: 30 cm. They had plenty of height. All the pipe ends were properly capped. He moved on to the next unit.

The hours went by. At one he went to the temporary QC office on the third floor. Mr. Pak and many of his workmen were there.

"Hello, Mr. Pak."

"Good afternoon, sir." Mr. Pak bowed. The workmen behind him rose to their feet.

"Ready for the quality control meeting?"

"Yes."

They all sat down. Mark and Mr. Pak on rickety folding chairs, the foremen cross-legged on the concrete floor.

"This is your meeting," Mark said. Mr. Pak began to laboriously read the plumbing section of the specifications in English. None of the foremen understood anything he was saying.

After five minutes, Mark interrupted. "Mr. Pak, let's change our plan, okay?"

"Yes, sir."

"I will read the important spec sections, you translate into Korean for the workmen, okay?"

Mr. Pak made furious notes.

Mark read excerpts from the materials section and the submittal requirements section. "Mr. Pak, shop drawing submittals, very important. You get shop drawing approval from IIAA before your men start work, okay?"

Mr. Pak agreed, still writing fast.

"What are you writing?"

"Meeting report, for Quality Control report."

"I see." Mark thought for a moment. "Okay, now I will read certain important parts of workmanship specification and you instruct the foremen in Korean, okay?"

Mark read the sentence from the spec requiring below-grade cast iron plumbing drain pipe to be stubbed up at least 15 cm above the top of the concrete floor slab. "Now you tell the foremen."

"Cast iron stub at least fifteen centi."

"No, in Korean."

Pak hesitated.

Mark pulled a yellow high lighter from Mr. Pak's shirt pocket and highlighted the sentence in the spec to Mr. Pak. "Now read this to them in Korean."

Pak spouted off a tirade in Korean with much arm waving.

Mark highlighted another sentence in another paragraph. Mr. Pak was trying to speak in Korean and write the sentence down in English at the same time.

"No need to write, Mr. Pak," Mark said. "I will highlight these words. You will read them in Korean to these foremen. Tonight you copy them into QC report, okay? Writing later."

Mr. Pak looked puzzled.

"I give you this paper. Tonight you put these words in your QC report, okay?" Mark pantomimed with high lighter, spec, and Pak's notebook. "Report later. Talking now."

"Okay, okay, Report later."

"Yes. Report writing later for QC report tomorrow."

"Okay, okay."

After that, the meeting went a little better. Mark called break so the foremen could stand up for a couple of minutes. They all stood talking and smoking Pine Tree cigarettes.

Mark finally closed the meeting at 3:00 and walked over to observe the concrete placement on the last parts of the floor slab in the mechanical rooms. Checkerboard pattern forms were laid out. Mark assured himself the underground pipe hubs were about 15 centimeters above where the concrete slab top would be. He checked the entire second floor. It was almost seven in the evening when he finally got in his Blazer, drove to his room, took a quick shower and fell immediately asleep.

He woke at five a.m. Outside, the desert dawn was pastel pink and orange, the mountains ochre. He grabbed a quick breakfast at the mess hall and was back on the job site by six a.m.

"Good morning, Mr. Pak," Mark said, returning the Korean's salute. "Let's finish the piping inspections, shall we?" They got out the drawings and worked together measuring the dimension, location, slopes and types of piping on the entire floor.

They finished at eleven and Mark drove to the Kendall mess hall for lunch.

Bill Vance, Dick Davis, Mike Robb and Ken

Cooley were there drinking coffee.

"Out in the desert, less than an hour's drive from here, there are lots of rock carvings," Mike Robb said.

"Which way?" Mark wolfed his sandwich down. He hadn't realized how hungry he was. "Straight out the *wadi*." Mike said. "Toward Syria."

A Palestinian mess boy tilted more coffee into Mark's cup.

"I'd like to see them. The sense of antiquity all around here," Mark said. "It's fascinating. Two of the ancient caravan roads cross here, the one from the Mediterranean going down the west coast of Arabia toward Yemen, the old kingdom of Sheba. You've heard of the queen of Sheba haven't you?" Dick grinned. "And the other road goes from Egypt East to Palmyra and through Iran to China. Marco Polo's Silk Road."

The three men gave him a politely disinterested look.

Suddenly there was an earsplitting roar, followed by a rumble that shook the sheet metal building. Mark scrambled out the front door with the rest of them. The mess boys were outside, staring at the sky.

Two French made Rafale fighter planes, flying low, were just disappearing over the hills to the north.

"Damned Israeli spies," Mike muttered.

"Checking up on their neighbors."

"In Saudi airspace?" Mark asked. They made their way back inside and resumed their lunch.

"They fly where they want." Ken said. "The Pentagon gives them our military satellite pictures, or they steal it from us, so they know how to avoid Arab air defense systems."

"I thought the Israelis were supposed to honor the peace accords Carter forced Begin and Sadat to sign."

Mike Robb laughed. "The Israelis have never let peace treaties stop them from doing what they want around here. Like making preemptive strikes."

"Including strikes on American forces if we get in their way," Ken said. "A buddy of mine was on the USS Liberty back in 1967, a U.S. Navy intelligence gathering ship in international waters off the coast of Lebanon. Israelis attacked it, killed 75 crew, almost sank the ship. Later they claimed it was a mistake. They hammered that ship for four hours – that ain't no mistake. They didn't want us seeing what they were doing to South Lebanon."

Dick signaled for more coffee. "I remember that incident. Captain of the Liberty got the Congressional medal of honor..."

"Yeah," Mike said in disgust. "But Congress was, and still is, so browbeaten by the Israeli lobby that they gave the medal to the captain in private, so they wouldn't have to explain to the American public that Israel attacked the United States and we did nothing about it."

"We've provided more foreign aid to Israel than all other countries combined," Mike continued, clearly angry. "And we get nothing in return. Except trouble."

"I gotta get back to the job," Ken said, eager to change the subject. He nervously eyed the Palestinian mess boys lounging against the back wall. Spies could be anywhere.

Mark climbed into the stifling Blazer and cranked up the engine, letting the air conditioning roar on high for a minute. "I wish there was somebody here who wanted to talk about something besides construction," he muttered to himself. "I need a vacation. That trip to Damascus Floyd was talking about sounds pretty good."

He drove to the IIAA office to check the dormitory project in-box and bring anything of interest to Redding. Redding's dusty Chevy was parked in front of the office. Inside, Jim sat in a guest chair at Ursula's desk, hard hat on his head, going through his in-box. "Ah, Mark, glad you're here." He tossed Mark a submittal. "Look this over and approve it, will you."

Mark started back out the door.

"We need it ASAP," Jim said. "Dae Joon's got their cast iron pipe on site. We need to get it installed and backfilled so we can schedule floor slab concrete."

Mark retreated to a table in the corner, fuming. He pulled a set of the project specs off the shelf, opened it to section 15000, then took a set of half-size drawings off the rack.

Redding gathered up his paperwork and went out.

After a few seconds, Mark closed the specs and went out to catch Redding before he left.

"This is not the right pipe." He squinted at Redding's sun glasses. Jim took them off and stared at Mark.

"It's service weight. Spec calls for heavy duty below grade for any building over two stories tall."

"Everybody else is using service weight," Redding's pale blue eyes bored into Mark's. "I'm not going to delay the whole job while they reorder, and probably hit us with a claim for the restocking fee, the shipping costs both ways, and an expediting fee to get different pipe air-freighted in, plus overall schedule impacts to the project. The Koreans may be polite, but they are sharp as nails and they are here to make a profit. They won't let this one go."

"Service weight doesn't meet spec."

"Will it work?" Jim said sharply.

Mark hesitated. "Yeah. But this sand is highly corrosive. Take a look at the galvanized pipe Holzmann used for water lines in this compound. Five years in the ground and they are rusted through."

"Our pipe is cast iron, not galvanized steel."

"If Dae Joon wants to use service weight, we need a credit change and an okay from tech section," Mark said defiantly.

"Ha Li used service weight at the Training Facility. Dae Joon told me so."

"That's Ray Barton's project."

"And the Dorm is mine." Redding slipped his sunglasses back on, "Get the service weight pipe approved ASAP." He got in his car and drove off.

Mark stood there sweating and furious, then got in his Blazer and drove around the compound for a while. I could take the easy way out, approve the submittal, let them install the pipe, nobody will know, he thought. Unless IIAA's home office happens to review the submittal in the file, which they might.

Mark pushed a cassette into the tape player, then punched it back out. No, I'm not going to give them a free pass on this one. The pipe they bought will work, but it cost them less than what they bid on the job. If I don't call them on it, the difference in cost is pure profit for them. If Preece were someone I could talk to, I'd ask him about it. But he'd screw me on this for sure if he got involved. I'll have to figure it out myself.

After a while Mark turned around and drove back to the IIAA office.

"I need to book a radio call to the home office," He told Ursula.

"Sure. I can get you on the schedule next Wednesday night."

"I'll write out the issue now, and you fax it to them today, that'll give them time to study it before we talk."

"One a.m. will be about eight a.m. their time." Ursula marked her appointment book. "That's when most people call."

Most evenings Hager lurked in his room, just sticking his head out, eel-like, from time to time. He'd caught Mark coming back from the mess hall.

"Not tonight. I don't feel like sitting around drinking bad beer with the same guys again tonight," Mark told Hager.

"There's a party at HSC tonight," Hager drove to the row of houses HSC had built for its married employees and found the house he was looking for.

"Tom's an x-ray technician at the HSC hospital," Danny told Mark.

Tom's wife Trish, a smiling chunky woman with streaked blond hair and a big grin opened the door. Her middle tooth was slightly chipped.

"Come in, come in," she said, motioning them inside. She led them to the living room, already full of people. Mark saw Ray and Sheila Barton standing at the side of the room and waved. The rest were hospital people he didn't know.

Drinks were handed around.

"I was selling Fuji x-ray film, the green boxes, at the time," Tom was saying. "Hardest work I've ever done in my life."

Trish came out of the kitchen carrying four glasses of home brew pressed to her chest. "The hardest work I've ever done was working on that fixer-upper house we bought in Long Beach," she said. She passed the beers out. Her tee shirt was wet from the glasses. Her big nipples showed through.

"You work your ass off to build up your contacts, to really get a client base, you know," Tom continued. He lit a cigarette and crossed his legs, fully self-assured as the center of attention. "Unless you've done it for a living, you can't appreciate the delicacy and the difficulty of selling something to someone. You have to be self-confident, friendly, helpful but not pushy. Knowledgeable for sure." He grinned his big Hemingway grin. "And willing to work sixteen-hour days and spend half your income on clothes. You want to look just right when you meet with clients, you know. You do all that, and it still may not be enough."

Mark noticed an interesting silent interplay was going on between Tom and Trish. Tom kept looking her way while she pointedly kept looking elsewhere. Tom hoisted his glass "Skoal. You work for months to get an account and to keep them happy, and then one day you come by and see yellow boxes, Kodak film, stacked in the corner. Loyalty doesn't exist anymore. Kodak, Fuji, it doesn't matter to your client. Whichever salesman has given them the best deal that week, that's who they buy from. You have to be tough to keep a smile on your face, keep sluggin' away in the face of that rejection."

Trish discreetly rolled her eyes.

"But it's a seductive way of life," Tom continued. "Kind of like gambling. One big sale will keep you trying for weeks of no sales. But I was good enough to pull down fifty K a year, which wasn't bad money back then." He stared at Trish's back. "Enough to buy that fixer-upper house in Long Beach, remember?"

She turned. "Yeah. I remember. We spent over a year fixing it up. And I still think we should have stayed there."

Tom shrugged the comment off. "We were doing something to the sheetrock on the bathroom wall one day, remember?" Tom laughed. "You were trying to use the sidegrinder."

She laughed and crossed her arms over her big boobs.

"The handle slipped out of your hands while the blade was stuck in the wall. So the handle's going around and around, tangled up in your tee shirt."

Tom was laughing so hard he could hardly speak. "Sheetrock dust was everywhere."

"That thing was beating the hell out of my left tit." Trish was laughing now. She held her breasts. "I was sore for a week."

Hager was giving Mark the high sign. "You ready to go?" he whispered.

"We just got here."

Hager squinted around the room, uneasy with people he considered his betters. "Let's get going."

"All right," Mark said.

In the car, Hager commented, "If I had a wife like Trish, I'd pay her more attention than Tom does. I hear through the grapevine she's not happy, wants to go back to the States. They'll probably get a divorce. Happens to half the couples that come over here."

Mark shook his head, "You must have spies everywhere."

Mark pulled a set of architectural half-size drawings off the rack and opened them to the third floor sheets. He sipped a cup of bitter ginseng tea and let his eyes move over the clean building lines, including the dimension lines and engineering lettering and numbers. The smooth, clean linearity of the drawings, thoughts captured, laid out on paper with a rectilinear beauty. Clean and self-contained.

I wish my life was like that sometimes, Mark thought. But life is more like the reality of construction, in which adjustments and compromises have to be made to make it work. No design is perfect; they all have flaws, and our job in the field is to make things fit together, on time and budget, so that the finished building works and looks perfect.

He turned the pages to the civil engineering sheets, the grading and backfill, trenching and below-grade piping. The drawings made everything seem neat and orderly. Trench sides were straight, lifts of fill were even and exact, final contour lines curved perfectly into place. But the reality was that errors and omissions would be found in the drawings. Arrangements could not physically be made as shown. There would be mismatches between where a concrete wall and a trench would be shown on different pages.

Mark thought, My job is to resolve and correct these problems in the construction phase, cost effectively – not compromising quality, but making things work that had been designed wrong while maintaining the intent of the design and providing a complete and usable facility.

Keeping quiet, he leaned back in his chair, careful to keep the drawings open. Couldn't have the Koreans thinking he was daydreaming. This'll be my sixth year with IIAA. Those first few years were tough, a lot to learn, the practical side of designing building mechanical systems, but it was fun. Now this year, seeing a different side of the work, the construction of the building and the mechanical systems in it. But it's fun too. Quality assurance in the field is fun. He stared at the drawings on the sheets in front of him. It's hard work, but fun work, taking a client's completely unformed ideas about how he wants a building to look and function, providing him a design that looks the way he wants it, is functional, inexpensive to build and operate, and meets all construction and safety codes. Getting all that down on paper, these beautiful design sheets. Then after the client has funded the project, and the home office has awarded the contract, establishing the schedule and budget and doing daily checks of progress and quality. The contractor comes to me to interpret the design, deciding what needs to be done to get it built right. I am the oracle at Delphi, he said to himself.

Mark grinned at the concrete structure taking shape out the window. He thought of major differences; unlike the oracle, his pronouncements would have to be perfectly clear, have to work every time, and have to keep the complex play of phases of work on schedule and on budget.

Decisions always had a cascading effect. A change here could affect everything else that came after, so interpretations and revisions of the design had to be fully thought out or the follow-on work would require a bigger, and more expensive, change order. Occasionally a change brought the whole project to a halt, while parts of the completed work were removed and redone. And in that case, the construction contractor would claim additional time and money was due him, and the government must pay it. Preece, Hager and Barton feared that above all. They were so afraid of making a mistake that they never made any decisions. Better to shift questions to somebody else, tell the contractor to go back and read his drawings and specs and contract clauses or, in other words, solve it himself. They were very quick to exercise their right to say something was wrong, but very reluctant to say when things were right.

Mark flipped forward a couple of sheets. The architectural sheets were the best. The smooth, perfect details of soffit and eave, window mullion, door jambs and framing, roofing details and the cabinetry in kitchens and bathrooms. The representation that was so much cleaner, clearer, neater than reality. Details as beautiful as illuminations in sixteenth-century manuscripts. He loved the drawings.

Mark got up and looked out the window at the job site, obscured for a moment by blowing dust. The wind boomed across the sheet metal roof of the temporary office. Now he was part of the construction process, the literal process of creation, assembling a building where none had been before. The sense of accomplishment was intoxicating. The dust devil passed and the gray pre-cast concrete walls of the Dormitory building materialized against the hazy white sky.

Lost again in thought, Mark's mind continued. All these workmen, me and the designers, the planners and organizers, all of us will soon disappear like ghosts. Only the building will remain. The occupants will move in, not having a conscious knowledge of the thousands of hours of work that have gone into making this clean, cool, finished product; doors and drapes that open and close smoothly, air conditioning that runs silently, clean water that flows from polished faucets, the bright lights all functioning perfectly. Fresh carpeting and tile on the floor, painted walls clean and bright – all with the smell and look of clean and new. By then, all of us construction ghosts will have moved on to the next project.

Redding appeared out of the blowing dust, banged in, and sat down at his desk. Mark closed the drawings and went back to his daily quality assurance report.

Chapter 5

Mark took two bottles of Fix beer and a bottle opener up to the roof garden of the hotel. He found a metal lawn chair near the clothes line, faced it away from the morning sun and propped his feet up on the parapet.

Before he'd left the airport, Mark had had the tourist desk book a room in a hotel near Syntagma Square; then took a taxi into town. With an hour to kill before Jennifer's flight arrived, Mark enjoyed the solitude of the hotel's inviting and empty garden. While he sipped his Fix beer, he admired the fine Mediterranean sunshine, casting its pale light on whitewashed apartment buildings and olive trees set against the rocky hills. He could see the famous Acropolis and the tall hill with the monastery on top; all around lay Athens. The light was beautiful. A smile crept over his face.

The sun and air were dry and warm, the smog gray over the city. Traffic noise reflected up the canyon of the street into the blue sky overhead. The panorama was spectacular, from the rocky hills past the Parthenon to the hint of glitter on the water at Pireaus harbor. Jennifer would be arriving at four p.m. on a TWA flight and Mark had been looking forward to seeing her ever since he stepped off his Gulf Air flight in Athens.

As she came through the customs line and the glass walls that separated the greeters from the incoming passengers, Mark caught glimpses of Jennifer's long black hair. She had a single piece of yellow Samsonite on a cart coming out the doorway. He watched her throwing occasional anxious glances toward the crowd while she cleared immigration and customs.

She beamed when she saw him waiting.

They rode the number 11 bus into town like old hands. Mark chattered about inconsequential things as they rode past the dusty olive green landscape, the buildings under construction and the auto repair shops. They got off at Syntagma square and walked the two blocks to the tiny Cleopatra Hotel.

The day and night passed like dreams in the warm still sunshine and it seemed like they had never been apart. In the early morning, Mark woke lying next to Jennifer, and silently looked at the view of the city outside the tall French doors, and wished he felt more. Homer's rosy-fingered dawn began to paint pastels across the distant hill and the tall apartment buildings.

They drank strong tea and ate hard rolls in the cafe downstairs, then set out for the day's exploring: the National Museum with its Mask of Agamemnon, a bus to Sounian Bay where they walked the temple of Poseidon, then back to Athens for a stop at the Acropolis. They walked back to the Plaka at dusk and stopped in a tourist restaurant at random. During their prix-fix dessert, dancers formed a line, danced round the room, and slung cheap plates into the fireplace already clogged with broken china. The over-touristy atmosphere of the place drove them out. They took a taxi to the hotel and drank a beer in the lobby, tired but full of the sights and sounds of Greece.

When they asked about going to Mykonos, Gregory, the desk man, said they could book passage to the island very cheaply. "It is almost Christmas. The boats are empty." He quickly produced two tickets on the passenger ship Leto, from Pireaus to Mykonos with short stops at Tinos and Artos.

Gregory was not happy about storing their luggage for three days. "I have too many baggages. Young people they go, say they will come back, but they never come back," He shrugged.

The shrieking of his Casio alarm watch woke Mark from a deep sleep at four the next morning. He clicked on the bare 40w bulb in the ceiling. They dressed, took a taxi to Pireaus, and boarded the boat. In their closet-sized cabin, they fell into a doze on the narrow bunks while the boat put to sea.

After a while, they roused themselves and had a cup of tea in the crowded ship's coffee shop. Rocky islands slid by on a dark blue sea. By mid afternoon, the Leto nosed past the breakwater and into the still water of Mykonos harbor. Open boats came out from the quay to meet the ship and transfer people and cargo.

Mykonos town rose whitewashed against the gray-green hills of the island. Dark rain clouds loomed up behind the rocky hills in other-worldly beauty.

Mark and Jennifer struggled down the ladder into the heaving boat and the vessel roared to the quay. They scrambled up the stone steps and stood staring at the travel-poster perfect beauty around them. The other passengers quickly collected their luggage and departed, and the square fell silent. The Leto was already moving toward the open sea.

Their guidebook recommended a harbor-front hotel, but they found all of them locked for the season. At the tourist police office, they were directed to a hotel up the hill. They slid into their backpacks and hiked up the gravel road as the wind freshened, carrying the scent of rain. Dark clouds blew over the ridge of hills behind the famous windmills.

The desk clerk seemed surprised to see them, but assured them he had "many rooms." They were shown to a cold room with tall windows overlooking Mykonos' town and harbor. Far out on the ocean horizon they could see the Leto, a dot of white against immensity.

They had tea in the lobby, then walked down the hill and located an Italian restaurant in the maze of sun bleached alleys. There was a ridiculous Greek musical comedy playing on the black-and-white TV perched above a row of potted ferns. The two waiters sat staring at it while Mark and Jennifer ate. They had finished their manicotti and were halfway through their second bottle of Domestica when three guys came in and sat down at the next table.

Early twenties, loud, obnoxious, obviously American, they were carrying on about how many traveler's checks each of them had left, and laughing over misspellings in the English menu.

Mark spoke up mostly to shut them up. "The manicotti's pretty good."

They settled down and placed their orders.

"Care to join us?" one of them said.

Mark traded glances with Jennifer, "Sure."

They rearranged themselves around the table and introduced themselves.

The tall, blonde-haired kid, David, was from Visalia, California, a guy with long black hair in a ponytail wearing a ski jacket was Bill from Oregon, and the other guy – a little older – was Paul, from St. Paul, Minnesota. They were all ex-teachers, recently quit. Jennifer brightened, while Mark resigned himself to the inevitable teacher talk, usually all complaints.

When the meal was over, the five of them wandered out into the windy night.

"If we're going to the Sundowner we go straight down this big alley to the harbor, then left toward where the lights are, then right past the big square," Mark offered. They followed him down the narrow alleyways.

The Sundowner was full of smoke and music and every tourist on Mykonos. But they all bellied up to the bar and shouted chit-chat at each other. Mark drank a Fix, and David was about to buy another round when Mark caught Jennifer's eye. "I think it's about time for us to be going."

Jennifer looked over the crowd in the smoky room, her long black hair hiding her eyes from him.

"Want to go with us tomorrow?" David asked. "We thought we'd drive over to Paradise Beach. We've got a rental Jeep and you're welcome to ride along. Going around nine."

"We'll meet you at the harbor square then." Mark said.

He and Jennifer strolled down to the edge of the harbor by the tiny Greek Orthodox Church where the night wind whistled over the weathered wall and the surf thundered against the breakwater below. The winter constellations were bright.

Mark's heart was clear and full of the same soaring love he had always had for her. They stood together in darkness at a forgotten church wall looking out over the surf-crusted blackness of the ocean.

He started to speak, something about the winter constellations, when she took his hand, came close to him, and whispered, "I really do love you, you know."

He put his arms around her and pulled her close but said nothing. Without a word they started back along the row of closed taverns and up to the stepped alley toward their rooms in the little hotel halfway up the hill.

They continued their way back to the hotel in silence. When they got to the front of the hotel, Mark rang the bell at the front door and an old woman dressed entirely in black came out of the room behind the desk, took their room key from its box and handed it to them, then returned to her room without a word.

The high ceilinged corridor was silent and cold. In

their room, Mark pulled the window closed and they undressed quickly and dived under the covers.

In the darkness, Jennifer's body felt warm and slim and familiar. They lay in each other's arms but did not make love. Outside, the night wind whistled and boomed.

Over the years, Mark was to remember this night as the last night he and Jennifer were truly together and things were just as they had always been.

The next morning they drank tea and ate bread in the quiet hotel dining room, saying nothing. Then they walked down the winding alleyway toward the harbor under an overcast sky. There was no wind and it was not cold. At the harbor, they stood looking at the water littered with detritus. The boats clacked and there was the sound of a broom as an old woman swept the sidewalk. Soon they heard a car coming up one of the narrow streets between the two-story whitewashed houses, and then the Jeep came into view.

Paul had the top down on the Jeep. They clambered over a tire and squeezed three into the back seat.

The drive over the hilly countryside was pleasant. Rock fences divided rolling green pastures. They got lost a few times on the narrow paved roads that wound along between the fences, but finally dropped down the last steep slope to the small beach that was called Paradise. It was empty of people in the winter season, although the brochures showed it packed with Scandinavian tourists in the summer.

The group scrambled out, scattering down the sand and up the smooth rocks at the end of the crescent. Mark and Jennifer wandered down the beach together, then climbed up on some boulders at one end and stared at the flat blue horizon. Islands were hazy in the distance. As the sun began to sheen through the overcast, the Aegean glittered. It was tranquil in the warming pale sunlight. Inland, there were rain clouds, but here it was warm and still.

They lay side by side on the sun heated granite glittering with mica. After a while Jennifer went down to the stony beach and walked along the edge of the water looking for shells and examining the polished stones, while Mark drifted off into a sleep troubled by anxious dreams.

When Mark woke, she was lying beside him again.

"There are no shells at all on this beach," she said. "Can you believe that?"

"I saw some graffiti that must date back to the sixties along the rocks down here and to the left," she said. "This must have been quite a gathering spot back in the Age of Aquarius."

She looked at him with sad eyes. "What's wrong between us?" She put her arms around him and started to say something more, but he kissed her instead.

They lounged back against the warm rock with their arms around each other, not speaking.

"I love you," she said.

"I love you, too," he said, knowing it was not true, but caring too much for her to say anything else.

They lay peacefully in the pale sunshine.

"A land that is always afternoon," Mark said.

"Lotus eaters," Jennifer said. "I sometimes wish…"

After a while she sat back up "We could walk back to Mykonos town," she suggested, "instead of riding with them."

Mark nodded and looked inland. "Yeah. I don't think it will rain until evening, so we still have time for a beer before we go."

They slid down the rocks and strolled along the beach to the little wooden taverna at the other end.

The Americans sat at one of the rickety wooden tables on the covered veranda that faced the beach. Three older Greek men were sitting at a table back of the bar, talking softly among themselves. Otherwise, it was silent.

"It's happy hour," David said as Mark and Jennifer pulled wooden chairs up to the table. "Sure would be nice to have a jukebox in this place. A little Foreigner, a little Journey, some Cars, Bee Gees."

"Bee Gees?" Paul snorted derisively.

An ancient Greek man shuffled over to the table and said something in Greek.

"Two Fix," Mark said.

The man stood blinking.

Paul said a few words in broken Greek and the man returned to the bar, got two bottles of beer from the shelf and brought them over.

"You speak some Greek?" Mark asked Paul.

"Just a few words."

They took some swigs of the beer, then Mark rose. "Jennifer and I are going to walk back to town."

"It looks like it might rain," Paul said.

"We'll see you back in town," Mark told them.

Jennifer picked up her beer and took another drink, then set it down without meeting Mark's eyes. She got to her feet and started toward the doorway toward the road back to town.

Mark caught up with her. "We can stay if you're not finished."

"No," she said. "I'm ready to go."

And they walked in silence up the steep hill. As they reached the top and stood getting their breath, the Jeep came flying up the hill, David driving, and roared past them with a blast of horn, then turned right at the fork in the road.

"I think I've offended you again," Mark said as they started walking.

Jennifer shook her head. "You never ask my opinion, just take it for granted that I want to do what you do," she said.

He started to remind her it had been her idea to walk back, but thought better of it.

"There are things we need to talk about," Mark said reluctantly. "But if I bring them up, it will ruin our day. Our vacation."

They walked down the quiet country lanes between low walls made of fieldstone. Forty minutes of walking put them on top of the hills above the town. Dark rain clouds were building in the western sky and the afternoon was rapidly getting darker.

"I love this kind of weather," Mark said. "Storm coming." They stood on the ridge line looking at the whitewashed buildings and the famous windmills of Mykonos town, brilliant white against the dark ocean. The wind was beginning to gust and he could smell the rain in the air. The pleasant fatigue of the long walk was delicious to him.

"We'd better hurry if we're going to make it to the hotel before the rain starts," she said.

He put his arm around her and turned her to him and kissed her. "I have never meant to hurt you."

"I know that," she said.

Halfway down the last bit of winding road to the hotel big drops of rain began to fall. A bolt of lightning flashed and thunder shook the air. As Mark and Jennifer came up the final hundred feet of alleyway, the rain began to fall heavily.

Inside the hotel lobby, they stood dripping and gasping for breath and laughing while the old woman got their room key for them.

Mark looked at his watch. "It's only four thirty," Mark laughed. "The rain's early."

"Lucky us," Jennifer laughed, combing her wet hair back.

The tall glass doors at the other side of the lobby framed the ghostly church of Mykonos town, partially dimmed from view by the blowing rain. At white table-clothed tables, seven or eight European tourists were having afternoon tea.

"After we get into dry clothes, let's change and have a nice hot cup of tea," Mark said.

In dry clothes, they took a table near the window of the restaurant, nodded to the other guests as they passed, and ordered afternoon tea. Outside, the rain pounded across the narrow tile balcony and dripped from the eaves.

The young girl who worked in the kitchen came and brought tea with lemon while they watched the rain sweeping across the terrace in gusts and flurries.

Jennifer poured more tea into both their cups and Mark was reminded of how sensual her movements were. Darkness was gathering, the serving girl turned on a couple of the low table lamps around the room. Later, in the darkness, they made love in the cold bedroom with the wind whispering and booming in the night.

After they'd dozed for a bit, Mark got up and got dressed, "I'm ready for some dinner…"

Jennifer dutifully began dressing, but it was clear she was not happy.

"I really don't understand why you have to go off by yourself for a year," Jennifer said, combing her hair.

"I've told you before, job opportunity."

"You sure that's all? You sure you aren't just looking for some reason to be away from me and from Missouri?"

"I do want to travel."

"I like travelling too," she said. "I love coming here, exploring Greece with you. Why can't we do it together?"

He shook his head. "It would get complicated if we tried to go to Saudi Arabia together."

"Why? You told me yourself there are couples there."

"I know," He shook his head. "I just want to take

this year by myself, try new things, be flexible..."

They both avoided the word marriage.

She continued with her questions, but now he was just avoiding them. When she started quoting him some feminist literature he began seething inside, but kept his silence.

"You just can't seem to be caring."

He wrestled his anger down. "I try to be caring."

"You have strange ways of showing your...caring," she said quickly.

"Question after question. Answering questions with questions. I'm no good at this." He took a deep breath and rolled his head on his neck. He had to get out of this stuffy room, away from these questions. "Let's walk down to the tavern, have a drink, eat dinner," he said.

She collected her coat from the tiny wardrobe and went down the corridor without waiting for him.

Outside, he could hear the wind murmuring around the eaves of the hotel and stared at the pool of yellow lamp light glinting off the shiny varnished wardrobe in the chilly room. He let his mind drift into an image from old movies, calendar pages fluttering off one after another as time passes. Our love affair is gone, he thought. Familiarity keeps us together, along with fear of loneliness, and the memory of how great things had once been. Like most of the married couples I see.

He locked the door, and entered the lobby where Jennifer was waiting, pretending to study the tourist brochures in the rack by the front desk.

The night was windy and very dark. They walked

quickly down the alleyways without talking. At the harbor they walked past a church, and paused at the wall overlooking the ocean. A hundred feet below, the Aegean was a tumble of waves on rocks. They peered inside the open door of the church, then stepped into the dim silence and took a seat in the last pew. Clusters of candles at both sides of the altar lit the sacraments and dim paintings of Balkan saints. After a moment they went back outside to lean against the stone wall above the ocean and looked up at the night sky cold and clear and full of stars.

They walked arm-in-arm up the alley to the restaurant.

In the restaurant, it was light, warm, and uncrowded.

He took her hand as they sat waiting at the table after they had ordered.

The fish dinner with calamari appetizer was quite good, but they did not talk much. Afterward they finished their wine and walked back through alleyways to their hotel. Inside, the room was still cold, so they undressed quickly, got in bed, and read for a while before they turned the lights off and lay quietly waiting for sleep to come.

Mark knew they would never make love again.

In the cold gray morning, they packed their bags before going down to a breakfast of tea and bread and eggs. They said nothing. There was nothing left to say.

Mark went back to the room to get their suitcases. The bags stood side by side against the wall. Outside the French doors, the day was the same as the day before – windy, cold, and gray. The ocean lay flat, its gray matching the atmosphere on shore. The Leto loomed over the breakwater.

Outside on the tiny terrace, the shutters rattled in the wind. Mark stared for a long time at the two suitcases standing against the wall.

It was all still there, the memories of the things they had done together, the days and years spent, all the way back to the day they had met. He could not imagine those years in any other way than with her. He thought about the night they had first made love, when they had been undergraduates at the University of Missouri. It had been in his friend Keith's tiny trailer bedroom after they'd been drinking wine and talking earnestly by candlelight. She'd said, "I love you, Mark. Let's make these moments last."

And for a handful of years, maybe his best years, they had made them last.

He turned his back on the wintery day outside, picked up the suitcases, and trudged down the corridor to the lobby. She was standing at the desk, wearing her familiar coat, her weight all on one leg the way she always stood. As he turned the last corner and came up the four steps to the lobby, she smiled a little at him and in that moment he knew in his heart they had lost each other forever.

Their last evening in Athens, they walked through the crowds in Syntagma Square to a little sidewalk café down a side street. Their conversation was minimal, polite, distant. Nothing about the future; nothing about the past. After a while, she fell silent and sat looking down at her half-eaten plate of spaghetti.

Her long black hair fell forward around her face. Mark touched her hand.

The next morning he rode the bus with her to the airport and waited until she boarded her flight. They hugged and said goodbye, but nothing more.

Then he took the bus back to the empty hotel room, went downstairs to the little shop, bought a couple of bottles of beer and took them up to his chair on the roof and drank them one after the other, staring at the hazy Athens skyline.

Chapter 6

Class convened at eight o'clock in the converted dining room of one of the IIAA leased houses in Riyadh. Fifteen IIAA employees from various sites around Saudi Arabia were seated on chairs that Mark knew he would hate by the end of the week.

The instructor stood at the front of the class, waiting for the clock to say eight o'clock exactly. He was a thin man with a neatly trimmed beard dressed in mall-new southern clothes. The instructor's resumé in the student notebooks said he was a senior scheduling engineer from the IIAA home office in Washington D.C. He had a southern accent you could cut with a knife. He was clearly nervous. The fifteen class members sat quietly. The older guys were clearly uncomfortable at being in a classroom and compensated for it by making sarcastic remarks. Even their body language said 'We know how things really work out in the field, and there's nothing you can teach us'.

The instructor began to describe how construction project scheduling was best done using a network of interconnected activities. This method had been used first in the late sixties for the U.S. Navy's Polaris missile submarine program, and later by the Air Force Systems Command to design, build and field

ICBMs. Now all major construction projects used network analysis scheduling, usually computerized.

"But the first step in developing a project's network schedule is to use pencil and paper," the instructor assured them. He pointed to a floor plan of a building tacked up on the side wall. "That's our project." Then he took his chalk and wrote five construction activities on the blackboard. "For a two thousand square meter building, tell me how long each of these should take."

"Four weeks for survey," somebody said. The instructor drew a straight line on the board and wrote thirty five under it. "Thirty five days."

"Two weeks for grading and trenching," somebody else called from the back of the room.

The instructor drew another line.

"Fabricate and set rebar, maybe four weeks..." another called out.

"But I'd start placing concrete as soon as the first foundation steel is set and the forms finished," came another response.

"Seven day cure?" the instructor asked and several people agreed. "So I'll draw me an arrow from this line, fab rebar, to this line, place concrete," the instructor said. "It won't go to the beginning because you just said good construction practice is to start concrete as soon as the first foundation forms are ready. So maybe the arrow comes in here three days after start of rebar fab and set."

He quickly linked the lines together with arrows and wrote numbers under each line, "These are the durations in days, including weekends."

A grizzled veteran raised his hand from the back row, just like in elementary school. And that was likely the last time he'd been in a classroom, Mark grinned.

"That's all nice and pretty, but it don't happen that way. There's always a delay, maybe not all the number three bar comes in on time so you've got to shift crews off to form work."

The instructor grinned, "That's right. There are delays and work-arounds. So how do you know what effect these up front delays may have on the project completion date?" He pointed to the distance past the right end of the blackboard. "Two years from now. Is it a day-for-day slip?"

"Nobody's going to give a contractor a day-for-day slip because of normal day-to-day glitches."

"You think he should keep the end-date fixed?" the instructor asked innocently. "How would you prove your case to the contractor?"

"Well..." the old guy wavered."

"Contractor will file a claim if you force him to speed up," the instructor continued. " All these Korean guys learned their trade working for American companies back in Korea. They know American procedures better than we do." There was a ripple of laughter.

"Well, that's the strength, one of the strengths," the instructor said, "of a network. You and the contractor can sit and analyze the delay activity and determine exactly what effect it will have on other activities and the overall project schedule. Look here. Some activities are going to be working in parallel with each other, but they don't all take the same amount of time. Therefore, some activities have 'float' which means excess time, spare time. And some activities don't have float. Once you've drawn your whole network, one of your first steps is to find the critical path for the project. That's the sequence of activities that has the least or zero float all the way through to the end of the job."

He traced a line in red chalk from activity to activity, left to right. Then he added up the durations of all the activities on this path. "Fifty-six days. That's your minimum project schedule for these activities. If activities off the critical path slip, you can still stay on schedule. But if the critical path activities slip, you have a day-for-day slip in the project completion date."

Next, the instructor showed several viewgraphs of a network from a large project in California. Mark sat entranced by the elegance of the lines and nodes, how the entire project could be organized and monitored so simply yet so effectively.

In the years to come, Mark would always remember that afternoon in the darkened classroom in Riyadh as the time he realized he wanted to spend his career managing large complex projects.

When the class ended at 4 p.m., everyone stood talking shop for a while then drifted off into the afternoon glare. The instructor was standing by the door uncertainly. "Where do you folks usually eat dinner?" he asked Mark. "Any restaurants nearby?"

"Yes, but most us eat at the IIAA cafeteria. It's over past Pepsi Circle."

"Very far?"

Mark grinned. "I'll drive you over there."

He sat white-knuckled as Mark nonchalantly hurled his Blazer through the slashing Riyadh traffic.

At the cafeteria they had a pleasant dinner of meatloaf, mashed potatoes, and green beans, with apple pie for dessert. Mark traded him some Saudi riyals for U.S. dollars so he could buy dinner. From his exhausted expression, Mark could tell jet lag had a grip on him, so he drove him back to the transient facility with minimal conversation.

The next morning's class was even better. Mark loved the clean linearity of the diagrams, neatly breaking a construction project into its hundreds of activities, connecting them to show their interdependence and how a delay in one would cause a delay in others.

"Obviously, if you add more crews to an activity, say placing foundation concrete, the task duration becomes shorter, but there's a point of diminishing returns," the instructor told them. "The job site will only accommodate so many men, so many concrete pumps, so many ready-mix trucks. And, of course, your cost goes up. The trick is to find the optimum cost point." He drew a chalk line down to the midpoint of the concrete-placing activity. "Here's your structural steel delivery. We don't have to wait until *all* the foundations are placed and cured, we can start setting steel as soon as the first foundations are cured. So then you've got both structural steel and concrete happening concurrently." He drew a long line to the left from the structural steel assembly activity. "You need to have ordered your steel three months before to allow for design and fabrication time at the supplier, plus however many months it takes to ship steel to the job site. So designing and ordering the steel may occur before you even break ground on the site."

That evening Jim Redding said he wanted to visit with some friends of his, but his car had a flat tire.

"You meet your friends here, I'll go get your car for you," Mark told him. The IIAA staff had towed it to the IIAA motor pool while Mark and Jim were in class.

Mark arranged to ride with Bill Caruthers from the IIAA Riyadh office out to the motor pool to retrieve Jim's car.

Brian had asked Mark to go with him to a party at one of the Lockheed compounds after dinner and Mark had reluctantly agreed to go. He checked his watch, there was plenty of time to get to the motor pool and get Jim's car and get back.

"We got an errand to do on the way," Bill told him as they got in his car. "Won't take long, we got to witness this guy being escorted out of jail by the Saudi police."

"What's he in for?"

"Selling liquor to some Filipinos. Not just once but several times. Trying to make money on the side."

"That sounds stupid," Mark said as they drove through the old part of Riyadh.

"You got that right. This guy has spent three days

in prison. Embassy's got his release and a ticket out."

There was a stink in the air.

"Damn!" Mark said. "What's that smell?"

"Jail," Bill said grimly. "The U.S. embassy and the Saudi government require prisoner release be witnessed. So that job falls on us. I hate it. We rotate it around all of us at the IIAA office here. This month it's my turn," Bill said. He parked the car in an alley, jumped out, slammed the car door and stumped off down the dusty alley. Mark hurried after him. "Let's get it over with."

When Mark caught up with him, he asked, "They hold executions inside?"

"Naw, they're public. Noon Fridays over at the square by the clock tower."

Around the corner was an open square with a cage of steel bars twenty feet tall. The cage was half a block long. Bill and Mark approached a crowd, mostly Filipinos, outside a barricade twenty feet away from the cage. Most had bandanas on their faces, bandit style. The stink was overpowering.

"It smells like shit."

"That's what it is. No toilets in the cages. Every couple of days the guards take fire hoses and hose down the prisoners and the cages. It runs down these streets." Mark stepped carefully. Flies were unusually thick, working the corners of any uncovered eyes and mouths. "Don't let these flies get in your mouth or eyes. Get some pretty nasty diseases." Bill slipped a dust mask on and handed one to Mark who hastily slipped it on.

A couple of Saudi guards unceremoniously shoved the crowd back from the barricade, then let them through. Bill and Mark followed the crowd.

The cage was divided into section, each full of men, mostly third world men – Filipinos, some Bangladeshis, lots of Sri Lankans, some Indonesians, a mix of other races. He saw two ragged Brits lying along the side of one of the cages.

The visitors lined themselves up against the waist-high fence set about two meters in front of the face of the cages. They laid bundles of food, cigarettes, and clothes on the dirty concrete.

The sound of coughing and jabbering in a dozen languages was a deafening din. A couple of guards walked down the row of offerings, probing them with nightsticks. Some they kicked to the prisoners; others they kicked back to the visitors.

Bill leaned close. "There he is." Four Saudi guards hauled a ragged guy out of the filth and marched past Bill and Mark and threw him in the back of a police wagon. Bill nodded and initialed a paper on a clipboard one guard with tribal marks on his cheeks shoved in his face. The police wagon drove off. Inside the cages, prisoners scuffled and tore at food and cigarettes within reach. A Brit fought up to the front of one of the cages shouting something unintelligible toward Mark. Bill pulled a pack of Marlboros out of his pocket and tossed it to him. The Brit had it but a dark hand from the crowd twisted it away. "Let's get the hell out of here," Bill said. They went back down the stinking street and rounded the corner.

"Hell of a place, ain't it?" Bill said. "No civil law

here, only Sharia, the law of the Koran. All proceedings are held in Arabic and the court ain't going to translate for you. There ain't no 'right to a speedy trial.' Hell, there ain't no trial, in our way of thinking, no appeals, no lawyers. If you's a prisoner, your friends have to bring you food and blankets – all's the prison provides is tap water, and they throw a pile of bread through the bars once a day." They got in the car and Bill turned the air conditioning up full blast. "You don't want to break the law here."

When Mark got back to the IIAA leased house, Redding was still sitting on the couch in the living room watching a videotape on the Betamax and reading the European edition of the Herald Tribune. Mark tossed him the car keys. "Thought you were going out tonight."

"Cancelled," Redding said. "Look at this." He held up the front page. The headline read: Shah leaves Iran. "That's going to be a problem," he said to nobody in particular. "Secretary of State Vance just sold him one hundred sixty more F-16s. Iran's been in turmoil for a year, riots and demonstrations against the Shah."

"Why's that?" Mark asked.

"Economy is in a shambles, despite the oil income, because of corruption in the government, wasteful modernization programs run by the Shah's cronies, and suppression of all opposition, which is mainly from the hard-line Muslim sects." Jim took a deep breath. "It's a mess. If he'd kept the economy running so that the middle class would prosper, there would be none of this."

"Who's in charge in Iran now that the Shah is gone?" Mark asked.

Jim thumbed through the paper, then shrugged. "Nobody seems to know."

Brian Zeller arrived. "The Shah is out of Iran." Mark told him. Zeller shrugged, "One dictator out, another dictator will come in. But he kept Iran our friend while he was there," Brian said

Redding snorted, "Countries can't be friends, only people can be friends." The comedy on the Betamax ended and Redding got up, ejected the videotape, sorted through the pile of videocassettes on the table, selected a different one and inserted it into the Betamax. This one was *Battlestar Galactica*.

"I'm going to take a quick shower before we leave," Brian told Mark.

Mark laughed, "Think you'll find a hot date over at Lockheed? That's even less likely than finding one at our compound."

Brian ignored him.

"I'm going to walk down to the cassette store," Mark said to Redding. "We need some new tunes for the drive back tomorrow."

He took a page from the discarded newspaper and used it to open the metal door latch. The leased house faced west. Even in the evening, the aluminum and opaque glass front door got as hot as a pan on a stove. Mark gingerly worked the latch open and stepped out into the oven-like tiled courtyard. It was like walking across the bottom of an empty swimming pool on a murderously hot day.

He made his way out the front gate and down the

dirty street to an even dirtier cassette store. Inside were piles of bootleg copies of cassettes in eight or ten different languages from a dozen different countries. All of them were illegally made in the Philippines. The Somali behind the counter ignored Mark. He ejected one tape and inserted another tape of whining Sri Lankan music in his cassette machine and sat immobile behind the counter.

Mark thumbed through the English language tapes. The covers in the cassettes were bad copy machine copies of record album jackets. Sometimes the song list had been retyped, often with hilarious misspellings. Mark picked a Lennon cassette, *Imagine*, Fleetwood Mac's *Rumours*, and a Beach Boys' *Greatest Hits*.

"*Ashra riyal*," the Somali said. He had two vertical parallel scars, very old ones, on each of his jet black cheeks. Tribal scars, Mark had been told, applied in childhood. Mark handed him a ten riyal note and walked back to the house down an empty sunblasted street.

In the living room, he opened a Moussy imitation beer and sat staring at *Battlestar Galactica*. The scene seemed to be a futuristic casino. But Mark's mind was on network scheduling.

Brian showed up smelling of cologne. "Let's go, I know Nigel from when he was with Kendall up at Tabuk. He puts on a hell of a party."

Brian found the Lockheed compound easily enough. "There it is." He pointed at the Lockheed logo on the compound wall. The gate stood open and the Saudi guard ignored him as he drove in. The Lockheed compound looked much the same as all the other foreign company compounds. Two dozen ranch style houses, a cafeteria, store, post office and swimming pool in the middle.

The party tonight was at house number 2500, which was not hard to find. The front door stood open to a racket of music and laughter, talk and light and cigarette smoke. They parked down the block and walked back to the house.

They edged into the living room, packed with men and hazed with cigarette smoke, the roar of drunken talk, and music from a cassette player. Brian nudged him and nodded toward three obviously British women clustered together near the kitchen door.

"Coming through, coming through." a tall man with black hair shouted. "That's Nigel," Brian said. Nigel plowed through the crowd holding a brass tray over his head with three bottles of *sidiki* on it, a clear vicious homemade spirit. A couple of Brits standing nearby grinned at Mark, "Only the finest spirits served here..."

"...distilled under the Queen's seal." They laughed.

Nigel topped off glasses in the packed living room and shouted over the racket, "Raise a glass! To the Queen!"

A great shout went up, "To the Queen!"

"And to Boxing Day," a guy near Mark shouted.

"What's that?"

"Boxing Day? The first weekday after Christmas. Celebrate the contributions made by service workers."

Mark looked puzzled.

"Postal workers and the like."

Mark struggled through the crowd to the kitchen hoping for some elbow room. I hate this, he muttered to himself. A room packed full of strangers, too noisy to make conversation, nothing in common anyway.

Unfortunately Brian had had the same idea and was standing in the kitchen pouring himself a drink.

"A few women here," Mark rolled his eyes, "and about a hundred men. The odds are not in your favor."

Two guys, very drunk, staggered into the kitchen and began to haphazardly mix themselves drinks out of the many half full bottles and Pepsi cans that covered the counter top. They sampled each bottle in turn, roaring with laughter.

"What's Lockheed doing in Saudi anyway?" Mark asked Brian.

"Maintenance on the F-16 fighters we sold the Saudis."

"Brits are doing it?"

"Yeah, the Saudis bid out the maintenance contract, and the British subsidiary of Lockheed got it because they have lower labor costs using UK labor."

Brian laughed at the two drunks both trying to squeeze through the kitchen doorway at once. "We sure live better than these guys. I guess IIAA ain't so bad,"

"You seldom say much good about it."

A thin woman in a green sleeveless dress came

into the kitchen. She had dull brown hair cut short, smeared lipstick, a cigarette in one hand. "Fix me a drink, love," She shoved an empty glass at Mark. He poured an inch of Ballantine scotch in it. "No ice, right?"

"So you're American, then?" Then she smiled and pushed right up against Mark pressing him to the counter. "I'm drunk" she said.

Mark, at a loss as to what to do, gave her a hug and she put her head on his shoulder briefly. "And I'm Lorna," she said slowly and carefully in a Midlands accent.

"Then I'm John Ridd," Mark joked.

She pulled back far enough to focus one eye on him. "Bloody hell," she said. "An American that can read." She puffed her cigarette, not noticing it was not lit. "You're not drinking."

"I don't see any beer,"

"Just a minute," she said and opened a cabinet, leaning over to give him a good view down the top of her dress. She had unremarkable breasts.

Brian winked at Mark and made his way back into the living room.

"Try this lot," she handed him one of the green bottles everyone used for home brew and leaned against him while he managed to wrestle the bottle open. Her haired smelled of cigarette smoke. The beer was not bad, different from the stuff at Tabuk.

"Your husband is with Lockheed, I assume?" Mark said.

Lorna gave him a look, "What's it to you?" She lunged forward and planted a wet kiss on his mouth.

125

Beer slopped against his shirt.

Mark pulled free, slugged down what little beer was left in his plastic cup and dodged out into the mob in the living room. Brain snagged his arm. "Where the hell are you going?" he shouted into Mark's ear. Somebody had KC and the Sunshine Band turned up full volume. "You're not going to pass that up are you?"

"I definitely am. Let's get out of here."

The scheduling class ended at noon Wednesday. Mark and Jim ate lunch at the IIAA cafeteria, gassed up the car, then set off toward Tabuk. The road lay before them endlessly straight across a featureless, flat desert. The hours passed; the white light shaded toward mauve as dusk fell.

"No trucks on the road, that's strange." Mark commented. "Lots of them pulled over, though." They passed little groupings of four and five trucks pulled together in the desert, their drivers clustered around campfires.

"Ramadan starts tomorrow," Redding said.

"All businesses are closed during the day, open at night, right?" Mark asked.

"Yeah, for twenty-eight days, one lunar month."

They drove into a plum-colored evening. The first stars appeared overhead. There were cars parked near rock outcroppings out in the desert. Mark could see the white thobes of Saudis sitting on top of the rocks.

"Waiting for the first view of the new moon," Jim told Mark. "That signals the beginning of Ramadan. When Mohammed had to get out of Mecca, he crossed the desert to Medina travelling at night, resting during the day, so the faithful do the same."

"You seem to know a little about Islam," Mark said.

Redding kept his eyes on the road. "No more or less than I know about Christianity or Judaism."

Mark grinned, "Not like our colleagues who know nothing about Islam." They laughed. "Like Ray Barton, Brian Zeller, and my buddy Ed Preece, the born-again Christians."

Redding snorted, "Keep this between you and me, but those guys with their closed minds really irritate me. They keep wanting for Ti and me to join their Sunday evening bible reading meetings which I refuse to do. They are so opposed to Islam and Judaism too. I'm not a bit religious, so I don't see any significant difference between the three middle east religions. They all have a heaven and hell, a single god, a holy book, rigid rules..."

"...and self-appointed holy men to interpret those rules." Mark added.

"All three of those religions came out of the same place," Redding waved at the desert rolling past. "They will tell you that this one is much older than that one, but the culture back then didn't change that fast, they came out of essentially the same cultural context."

They drove in silence for a while, lost in thought. "I find it hard to believe they can justify all the wars, the pain, the suffering, they have caused over the centuries," Mark said.

"And U.S. support of Israel sustains that endless hostility," Redding said bitterly. The Carter administration likes being able to use Israel to channel U.S. military supplies to countries that the U.S. congress has banned sales to. It's a convenient way of for the President to circumvent congress."

"Like nuclear weapons technology to South Africa, right?" Mark chimed in.

"Right," Jim said. "Damned Israeli spies stole it from the U.S. for their own weapons program, now they are selling it on the open market. And we let them get away with it."

"Why do we let this happen?"

"The strength of the Israeli lobby. Congressmen know it's political suicide to vote against legislation AIPAC supports. And the fact that most American voters are incredibly naive. If they've been overseas at all, it's five days in Europe on a tour bus. I'm amazed the Saudis tolerate us at all. The Saudis see the U.S. supporting Israel against all logic and common sense and wonder why we continue to do it year after year. What do we get out of it?"

Redding glanced around at Mark, "Sorry about the tirade."

Back in Tabuk, Mark showered and changed, then drove over to HSC hospital. In pediatrics he found Cathy and Mary sitting in the lounge.

"No customers today?" Mark said. Cathy looked up and a huge smile spread over her face. Mary didn't look quite so enthusiastic.

"I just got back to Tabuk," he said, "Tony Cross is

128

hosting a party at the pool in our compound tonight and I wondered if you would like to go." He looked at both of them to be sure Mary understood it included her, too. They exchanged looks.

"I'd love to," Cathy said slowly.

"I'll pass, thank you," Mary said. She got to her feet and went to the nurse station in the other room.

"I'll meet you out in front after I change clothes," Cathy said, and hurried off.

Mark hesitated, looking at Mary's back. He started to say something, then turned and went back to his Blazer.

When they got to the party, there were about twenty people around the pool. The blue lights and the pool lights provided a cool counterpoint to the warm evening air. The sky overhead was royal blue.

"Tell me about your trip," Cathy said. They sat down in white plastic chairs facing the pool.

"Well, I spent almost all my time in class, but what I saw of the city looked a lot like Jidda, a mix of dusty sixth century desert town and ultramodern oil-money city, new cars everywhere, new stores with gaudy neon signs next to dirty shops, a mix."

"You two look cozy," Tony said. He had a tray with glasses and bottles of home brew and was making his rounds filling glasses. "Very domestic."

Cathy blushed.

Tony expertly poured two glasses full and handed them over. "I assume you know everyone here?"

Mark snorted at the ancient joke. "All too well."

Dick put a big-band tape into the cassette and a chorus of groans went up. Mark and Cathy circulated, chatted with people for a while, then slipped out the side gate closing it gently behind them.

The night air was delicious, the concrete block walls radiated heat from the day just ended. Arms around each other, they strolled down the row of tamarisks while the music and the laughter from the pool faded and only the soft sighing of the trees was left.

"Same old people, same old conversations," Mark said as they strolled, "No wonder we get on each other's nerves, seeing each other, and only each other, every day, at work and after hours, too."

"People sure seem to take a lot of interest in what you and I are doing."

Mark kissed her, "They are jealous of me."

"Brian is married, isn't he?"

"His wife left him a couple of months ago. He's pretending to be happy about it, but he's lonely as hell."

They walked in silence for a time. Then Mark stopped and kissed her, "We're lucky, Cathy, we're together for only one reason. Because we like each other's company." They walked down the sidewalk between the row of window air conditioners and went into his room. He left the light off and led Cathy straight into the bedroom. They made love slowly in the cool dimness. As Mark drifted toward sleep he thought he heard Cathy whisper something, but he was asleep before he could ask what she had said.

They woke about eleven, dressed and Mark drove her back to the HSC compound. "Can I see you tomorrow evening?"

"What time?"

"Let's have dinner at my room. I'll pick you up at six."

The next day Mark felt buoyant, tingling with life, a smile on his face and laughter ready to burst out at any time. He couldn't keep the grin off his face.

After work he walked quickly down the walkway between the roaring window air conditioners, eager to shower and go pick up Cathy.

Danny's room door stood open. "Hey, Mark!" Danny yelled from his cave-like room. Mark reluctantly stepped into the gloom. The usual guys, Dick and Brian and Ray were perched on their usual chairs.

"Get in here – it's happy hour." Danny meticulously poured home brew down the side of a plastic glass. Mark sat down on a rusty chrome chair from some discarded dinette set. "Things going good, ain't they?" Hager said to nobody and everybody. Dick smiled his scintillating smile from behind cigarette smoke. His hand caressed a glass of scotch. Danny eased himself down onto a recliner covered with a tatty red Indian blanket. Mark slugged down half his glass of home brew, "You know, every once in a while, I think this stuff tastes good. But this is not one of those whiles." He stood.

"No need to rush off," Danny said.

"He's got other interests now," Dick laughed.

"Yeah. You just need to be cautious about that stuff," Danny said, looking at Dick, not Mark.

"Why's that?" Mark asked.

"The women here can get you into trouble. The

Saudis don't like us fraternizing too much."

Mark set the plastic cup on the coffee table. "I'll keep that in mind."

Mark drove to the HSC compound and since he was early decided to take a look in the hospital pediatrics center to see if Cathy was still there. As he was walking up the sidewalk, he saw a big, young American guy standing in front of the rec center hands in his jean pockets, staring at the horizon. He must be new, Mark thought. I haven't seen him around.

Cathy was not in pediatrics, "She went to meet you, love," Elspeth told him.

When Mark came back out of the hospital and crossed to the dorm, the American was still standing there. Mark went over to him and shook hands, "I'm Mark Exner, with IIAA, don't think I've seen you around here."

"Dale Parker," the kid gave Mark a bone-crushing handshake. "I'm the new rec center manager for HSC." His accent and his manner was Midwestern.

"You from the Midwest?"

"Nebraska," he grinned sheepishly.

"I'm from Missouri," Mark said.

"University of Missouri. Tiger football. Yeah," Dale mused. "We play you guys most seasons."

"And beat us most seasons," Mark hurried to change the subject. "Been here long?"

"Almost a month. Not much to do, is there?" Dale looked lost. Mark knew the HSC employees were not issued company cars so they were stuck on the compound unless they bought their own cars.

"Tell you what, let's go diving at the beach one

weekend," Mark said on impulse. "I'll drive."

"That would be great!" Dale said. "Come over the rec center sometime, I'll play you a game of racquet-ball."

"OK," Mark checked his watch. It was almost six. "I've got to get going. Talk to you later."

Chapter 7

The Wednesday night poker game had become a fixture in Mark's routine. Every week he resolved not to go, but then changed his mind. He never had cared much about playing cards, but...

"You must really miss the scotch, huh, Kurt?" Brian said, eying Kurt's large glass of scotch on the rocks.

Kurt took an appreciative slurp. "I can buy it on the black market, Ballantine's scotch, for three hundred riyals a bottle."

Brian gave Mark a look across the poker table meaning: You can buy it, but you don't buy it when you can come over here and drink it for free.

Mark grinned.

"How about a cigar, too?" Brian said offering Kurt a cigar.

Kurt ignored the sarcasm. "Don't smoke. That's one vice I don't have."

"All those years in Mexico and so few vices?" Dick snorted. He threw a five riyal chip in the pot.

"Not so many years in Mexico," Kurt said. He folded his cards and topped up his glass from the bottle of Johnny Walker Red. "We used to spend the winter in Acapulco waiting for the spring break up in

Canada where I was working for an oil exploration crew." He sipped his scotch and studied his cards. "This was back in the late forties, early fifties. I came over from Germany in nineteen forty-eight."

"Didn't you tell me you flew for the Luftwaffe?" Dick slurred.

Kurt shot Dick a look that he didn't notice as he tossed a five riyal chip in the pot without even looking at his cards.

"Seventy-eight flight hours, most of it training; in Bavaria, in February and March nineteen forty-five."

"Trust you to be precise about the details, Kurt," Redding said with a laugh.

"For my efforts, I got to spend a year in a POW camp," Kurt continued. "But that was okay. Not far from Munchen. I worked on a gang hauling debris from bombed buildings during the day. At night I studied civil engineering and English. The Brits and the Americans were holding evening classes just to have something to do. So when repatriation came, I went to Manchester, England with one of the Brits I'd gotten to know. He helped me get a job with a mining company doing survey work in Canada, the Rocky Mountains in northern British Columbia."

Brian folded his cards and got up to top off the scotch drinkers' glasses. He took Mark's beer mug and brought it back full. Mark took a sip and found it was Heineken, not home brew. Brian made a sign not to say anything. Real beer was precious.

"So every autumn we'd drive down through the United States to Mexico to spend the winter in Acapulco. The company laid us off, but we had plenty of money. Up north we worked seven days a week and lived in camps out in nowhere, so nothing to spend it on. In Acapulco, back then we could eat veal cutlet, two-egg salad and a bottle of Bohemia beer for ten pesos. Served to you at a table under an umbrella right on the beach. Or red snapper, or a whole plate of shrimp, just ten pesos."

"You're making my mouth water," Vance said. "Who's bet?"

"Mine." Kurt threw in ten riyals, laid down a full house and swept up the pot. "Did that for four or five years. But the best times came later, maybe nineteen sixty-one, sixty-two."

"Let's be precise, Kurt." Brian jibed.

"I can tell you. It was October, 1962. Here's why I know." Kurt ignored the sarcasm. "By then I had gotten my papers, Canadian citizen, but I wanted something new, some different work. The American space program was just getting started. I had some money saved, so I bought a brand new red 1962 Corvette, drove across the bridge from Windsor, Ontario into Detroit and kept on going, all the way to Cocoa Beach, Florida. All the big American construction companies were setting up shop there. Two days later, I was working for Pan Am Construction Management doing the civil engineering for telemetry stations in the Eastern Test Range." He grinned his big square-toothed grin. "Those were great days. The whole area was booming. Construction was going seven days a week, sometimes two, even three shifts, building the housing at the Cape, launch pads, all the infrastructure for the Air Force and NASA. Everybody was working ten to twelve hours a day, seven days a week. I was living at the Palms Apartments. They raised the rent fifty dollars a month every month, there was such a shortage. Some guys were sleeping in their cars on the beach until the cops put a stop to it."

"No time for women," Brian dealt the cards.

"Lots of women. The bar at Ramon's or Bernard's Surf was full of women wanting to meet guys, every night. Buy them a surf-and-turf dinner and spend the night together. The strip joints were packed all night, go-go girls at the Caravelle lounge, the Sea Missile, the Pink Pussycat." Kurt clamped his mouth shut, realizing he was talking too much.

"Construction's like that," Dick said. "Boom days are fun days. Then bust. But if you're willing to travel, it's usually booming somewhere in the world."

Kurt finished his scotch, clearly embarrassed. "I need to be going."

"No need to rush off," Brian said. "You don't have all our money yet."

"Thank you for the drinks," He said formally and disappeared into the hot night.

Mark pushed his chair back. "I need to get going too."

Ed snickered. "Family man."

Mark grinned. "Yeah, but not this time. I've got to make a radio call to Washington, the home office." He checked his watch. "In fifteen minutes. See you guys Saturday."

"You going somewhere over the weekend?" Brian

asked.

"Oh, yeah," Mark said. "I forgot to get your permission. I'm driving down to Jidda Thursday, coming back Friday. Is that okay with you, Dad? Can I borrow the car?"

"She going with you?" Brian leered.

"Thanks for the beer, Brian. See you."

Mark closed the door and walked down the row of tamarisks whispering in the night breeze. He could hear Kurt's car making its way to the Kendall company housing area. He got into his Blazer and drove to the IIAA office. He let himself in, turned on the light and went into the radio room and flipped the switches Ursula had showed him. He sat waiting until the appointed time, when the relay radio operator in Jidda would patch him through to IIAA's office in Washington.

Construction is always booming somewhere, Kurt and Dick said. They'd spent their lives following those booms, around the world and back. The locations changed, but the processes stayed the same. These guys crossed and recrossed each others' paths over the course of a lifetime. No wonder they were always talking about the good old days in Pakistan or Guam or Germany or Venezuela, and telling all those anecdotes about people they all knew way back when. But after all those years on the road, what have you got? Where's home? Where do you go back to?

The room was hot with the big shortwave radio on. Mark opened his copy of the specs and drawings on the table, waited until exactly 1 a.m. then pushed the button on the microphone. The speaker crackled,

"Mel Adams here, Mark. I've looked over your request and we're agreeable to allowing service weight below grade. We'll get it to you in writing in the next day or two."

"How's the weather there?" Mark said.

"You guys always ask that," Mel said. "It was snowing when I drove to work an hour ago. Supposed to clear around noon, then get colder."

Mark sat staring at the radio dials, imagining a snowy Virginia day.

"I miss that cold weather."

"You wouldn't if you had to drive in it every day."

"Well, yeah. Thanks for your help, Mel."

"Anything else?"

"No. We going to see you guys on your annual tour of overseas sites soon?" Mark asked.

"Not until February. We've got to do Germany first."

Mark signed off, switched the radio to standby as per the typed directions taped over the keyboard, locked the office and drove home.

After lunch Thursday, Mark locked the house while Cathy loaded the British cooler into the car with some food, three cans of Fanta soda, and two bottles of Sohat water.

"All set?" he asked.

She nodded and they drove out to the IIAA maintenance shop and filled the Blazer's tanks with gas, then they drove over the dozen speed bumps and out onto the main road south.

He held his speed to about 120 kilometers per hour, about eight five miles per hour. The occasional car and Toyota pick-up passed him. He flashed around slow-moving trucks stacked precariously high with boxed cargo.

Cathy was all smiles this morning, as excited as a kid on her birthday. "I can't wait to see Jidda."

A Toyota driven by a bedouin kid roared around them in the face of an oncoming truck.

"If we live that long," Cathy said.

An hour later they had passed out of the flinty mountains with their veins of magenta, olive and pale purple minerals, and out onto the sun-blasted central Saudi Arabian plain, the Hijaz.

They talked for a while, then settled into silence. Outside, beyond their sunglasses, the air-conditioned air, the tinted glass of the windshield, the desert lay timeless, silent, and unforgiving.

They roared south down the arrow-straight road.

As the hours drifted by, the afternoon softened to dusk.

In the hazy flat distance they could see a back-up of cars and there were a couple of trucks pulled off onto the desert beside the road. Tail lights glowed pink in the beige dusk.

"An accident?"

"Don't know."

They slowed the Blazer into the line, keeping right at the rear bumper of the white Buick ahead of them to discourage line-breakers. A couple of cars behind them were already nosing out, looking to go around.

Typical, Mark thought, nobody knows what we

are in line for, but they all want to be first.

A truck rumbled past going the other direction and Mark nosed out to look ahead.

"Ah, I see it. Police and National Guard. Get out your ID, your passport, and your permission to travel form."

Cathy dug the papers out of her overnight bag.

They inched forward in the long line of cars. Young soldiers in new khakis with Czech assault rifles slung over their shoulders were checking paperwork.

"On second thought, put your papers away."

Mark eased the Blazer forward and ran the power window down. A soldier looked into the Blazer. The hot air from outside brought in the smell of his cologne. Mark handed him his passport.

"Hello," Mark said pleasantly.

The kid said nothing, flipped through his passport to the current multiple entry visa and studied it a moment.

"*Wayn ruha?*" he asked.

"*Ihua raayihiyn Jidda,*" Mark responded.

"*Nimayn jaay?*"

"*Tabuk.*"

The guard opened his hand and Mark took his passport back. The kid made a small brushing gesture with his right hand as he walked back to the next car in line, meaning they were free to go.

"What a jerk," Cathy said. "So impolite."

"Not really. He didn't think he was being impolite at all. His gestures are completely polite and appropriate in his society. Gestures and expressions are cultural."

They accelerated back up to 120 kilometers per hour on the open road.

"I'm glad you knew what he was talking about," Cathy said. "I sure didn't."

"He just asked where we were going and where we were coming from. Also it helps to have this red passport. They think it means I'm somebody important. It doesn't, of course. It just means I'm an employee of the government."

"He didn't even check my passport."

"That's another nice thing about this culture. They don't pry into what they perceive to be family units. Men don't talk directly to women they don't know."

They drove on into the darkness.

Another hour passed. There was literally nothing to see. The sky was hazy with desert dust, there were no signs on the road, no glow of the lights of a town on the horizon. The desert was dark and featureless on all sides. The only light and movement came from the occasional lights of cars going in the other direction.

The distant lights of an approaching car blinked on and off.

"What the hell?" Mark sub vocalized. He eased back on the gas pedal, his speed dropping to 110 kilometers per hour. The approaching car's lights grew.

Suddenly, the gray shape of a Mercedes truck with no lights appeared at the limits of the Blazer's headlight pools. Mark whipped the wheel left and floored the gas pedal. They dodged into the oncoming lights; the truck they were passing loomed alongside for an instant, then disappeared into the darkness behind them. Mark whipped the Blazer back into their lane. The oncoming car flashed past with a blare of horn.

"Stupid son of a bitch," Mark muttered, his heart racing.

"Those moronic truck drivers, driving at 20 kilometers per hour with no lights. They ought to be locked up," Cathy said. "Scared the hell out of me."

"Me too," Mark said.

They settled back into silence. Mark considered trying the radio. You could sometimes get the Soviet English-language news service which provided a unique look at world events. But it was unlikely. A pool of light on the horizon slowly resolved itself into a constellation of 40-meter-tall streetlights over an intersection in the road ahead. Fifteen trucks had pulled off the road and stood parked at random across the desert on either side of the road.

"What is it?" Cathy asked softly.

Mark slowed the Blazer. Ahead the road branched. Down one road were big black-and-white signs. He turned that way. As they approached the signs, he could see that he had made a mistake. The signs advised drivers in Arabic, English, Urdu, Hindi, German and Korean to turn back unless you were Muslim.

"Damn. Wrong way," he slowed and studied the tire tracks in the sand beside the pavement, then turned out across the desert, back to the other branch of the road.

"Where does that road go?" Cathy asked.

"To Mecca. The Holy City," Mark grunted.

"Only Muslims can go there?"

"Yeah."

He cut across the sand and got the Blazer on the right road. Eventually the glow of the distant lights of Jidda began to heave itself up over the horizon.

Mark turned the Blazer down Medina Road.

"Now. I need to find…"

"Can I help? Cathy asked.

"Look for a traffic circle with a big sculpture of a *dhow* in the middle."

He tried to get over but he was locked in a fast-moving stream of traffic. Unlighted side roads flashed by too fast to brake and turn into. Finally the lights of a big construction site illuminated a road.

Mark braked and turned in; cars roared past, honking their horns.

"Hey, this is the new Haj terminal." He paused the Blazer just short of the job site guard box. A Pakistani gate guard eyed him suspiciously. Under the Bechtel name and logo on the sign, an artist's rendering showed a huge series of white tent-like structures.

"Kind of a strange-looking building," Cathy ventured.

"I read a little blurb about it in one of the engineering magazines. Bechtel is doing the expansion project. Those are fabric roofs. Just like tents. Supposed to be reminiscent of the old days, I suppose, when pilgrims stayed in real tents on their way to Mecca."

Every intersection was like playing chicken. You couldn't hesitate or you'd get run over. He finally found the traffic circle with the ship sculpture in the middle, a white Buick that had been riding his rear

144

bumper swerved around them. A Chevy was cutting the corner from one radial street to the next. The Buick locked its brakes up and plowed into the right rear of the Chevy.

Mark snapped a glance into his rearview, whipped the Blazer to the right, then back to the left as the two collapsed cars spun around their mutual center of rotation twice and came to a halt against the curb.

Mark drove slowly on, turning down the street away from the wreck. A pickup, a Chevy Suburban, and a Toyota were stopping at the wreck.

"My God," Cathy said, peering over the seat back. "Should we stop?"

"No way," Mark said, driving carefully down the dimly lit street. An occasional working street light showed only closed gates and walled residences. They worked their way back to the main road.

"At a car accident, the cops take everybody at the scene of an accident to jail until they sort out who's guilty. And foreigners are usually suspects. If anybody is hurt or killed, somebody has to pay blood money to the victim's family. Being a woman means you are automatically innocent, but I have no desire to spend a night or two in a Saudi jail until IIAA can bail me out. I'm told that the prosecutors' logic in traffic cases here is that the accident would not have happened if you had not been here, in their country, therefore, you are guilty."

He realized he was chattering and stopped talking.

They drove on in silence. Cathy leaned against him as they drove.

After a while they found the IIAA transient facility and checked into the motel-like structure. Their room was big and new, cool and clean.

"I'm not tired yet. Let's go down to the gold souk," Mark said. "It's mind-boggling."

Gold. Stall after stall dripped with the bright metal. A thousand shapes of jewelry. Mark and Cathy walked down one passageway and up the next, past row after row of brilliantly lit glass cases.

"Look at these twenty dollar gold pieces," Cathy clapped her hands with joy.

The Saudi in the stall got out a handful of gold coins – U.S. twenty dollar gold pieces, Mexican double eagles, golden Maria Teresa dollars. Their weight and color and texture provoked an almost sexual thrill. Cathy's hair was the same color as the golden halo all around them.

"Are these real?" she asked.

"It's real twenty four karat gold," Mark said. "But counterfeit coins. All this stuff is made by artisans in Italy, I'm told. The coins don't come from any national mint."

She turned the coins over one by one.

"Beautiful," Mark said. He hefted the Mexican double eagle. "Kind of gives you an insight into what money was like a hundred years ago." He tried to flip the coin, but it was so heavy it didn't spin, just wobbled awkwardly in the air. "Or a thousand years ago when arab sailors raided and traded the Indian Ocean, and up and down the coasts of Asia and Africa."

"How do you know how much to pay for these coins?" Cathy said. "Twenty dollars?"

The Saudi placed the coin on the electronic scale and turned the digital readout toward her.

"That's the weight," Mark told Cathy.

Then the Saudi punched a number into a Seiko calculator and showed that to them.

"Weight times 24 riyals per gram, which is the price of gold today on the open market. So he'll sell us this double eagle for four hundred riyals."

She opened her purse and got out four of the big blue 100 riyal notes. The Saudi wrapped the coin in cotton and put it in a cheap brown paper bag.

"This is for you," she beamed, handing Mark the tiny bag.

"Really? Thanks." He got the Mexican double eagle out of the bag and admired it, mostly for her benefit.

The gold market filled Mark's senses with saturated golden sunlight, the heat from the display lights, the close smells of desert dust, diesel fumes, expensive European perfume, the Saudi women dressed in black abbayahs and veils, the smell of rotting garbage between two gold merchants' stalls, the sweat from two Yemenese laborers hand-mixing concrete.

The heavy coin glittered in his hand. He slipped it into his pocket.

"I'd give you a hug and a kiss," Mark said, "but since that's illegal here, I'll wait till we get back to the Transient Facility."

"That's not all I want when we get back." Her smile was dazzling. They walked on through the golden evening.

After a while, they came out of the souk and turned down the street where Mark's Blazer was parked. A merchant was pulling his storefront doors down with a roar.

"Must be evening prayers." They walked on into silence under the rosy glow of city lights reflected on hazy night air.

More merchants rolled down corrugated iron security covers on their store fronts. Mark and Cathy paused. Except that the women wore black abbayahs and the men were dressed in white thobes and red and white ghutrahs, it could have been a street in the States. The cars parked at the curb were late model Chevys and white Toyota pickup trucks. The street-lights were on, the wide sidewalks had been swept clean by the ever present Yemenese laborers in their orange jumpsuits.

"The Saudis are making progress," Mark said, "And here there's no street crime, no beggars." He could tell she was more interested in getting back to their room and admiring the gold she had bought. He touched her hair, "You sure look nice tonight."

She gave him a radiant smile.

"Other people will never understand all this," Mark waved his hand at the buildings and the street, the people passing. "How it feels, how it looks and smells. You can never really describe these experiences to friends back in the States. Unless they've lived overseas they don't really understand what you're trying to say." He looked up at the orange glow in the dusty night sky. "They hear what you're saying. Don't get me wrong – they're smart people, but they haven't been here. They haven't experienced it. Haven't smelled the air like it is right here, this moment - diesel fumes, that orange peel there against the wall, the distant smell of sewage, that French perfume you're wearing…"

"*J'aime.*"

"Yeah. The dust in the air, the scent of the bad stucco on this wall." He touched the wall. "You understand, I understand, but no one else ever will."

Mark parked his Blazer at the IIAA transient facility, a two story motel-like structure, shaded by tamarisks. He showed his ID and got his key from the Palestinian at the desk and they went up to their room where they undressed quickly, crawled into the clean new sheets and made love, Cathy gasping and crying out, Mark feeling his heart would burst.

Afterward they lay in the soft light, letting their breathing return to normal.

"That was wonderful," she whispered.

He nodded. "Yes, it was." He kissed her and rolled over to rest. After a while he heard her say very softly, "I love you, Mark."

He pretended to be asleep, and after a while he did sleep.

Sunrise was coloring the room when he woke. The sun was turning the room too warm. He rolled the blanket back. Cathy stirred and turned toward him, but he turned away and went back into a pleasant unmemorable dream. Cathy lay on her side, the sheet tucked neatly under her arm. In sleep, her face lay relaxed, mouth closed, the strain-lines around her eyes smoothed out.

After a while, they showered and packed their things. Outside on the balcony, the air was velvety and the sunlight on the buildings was a palette of pastel pinks, whites and yellows.

"I love this time of day," Mark said.

There was almost no traffic on the streets so they got to the IIAA Cafeteria in five minutes. The parking lot was full of Chevys.

"Popular place on Friday morning," Mark commented. The sign at the door said breakfast buffet, ten riyals. "That's why everybody's here," Mark said. "Americans never seem to pass up a deal on cheap food."

They loaded their trays with eggs and bacon, along with toast with jam from Denmark. Cathy got a small waffle with maple syrup. They found a table in the crowded room.

"This is real bacon!" Cathy exclaimed.

"Yeah, IIAA can get a limited amount of pork products as long as we only use it among ourselves."

A big man and his wife on their way out detoured to their table. "Thought I saw you here. How are you getting along?" It was Ron and Laura Stevens.

"I'm doing fine. How about you guys?"

Ron grinned and shrugged, "Life in the desert."

"This is a friend of mine, Cathy Locke," Mark said. They shook hands. "Ron was my sponsor when I first got here." Ron and Mark both laughed at the memory. "You sure were a welcome sight that night at Jidda Airport when I first arrived," Mark said.

"And now you're an old hand too," Ron said. He turned to Cathy. "You should have seen this guy's face," Jim said. "I thought he was going to tell me he was leaving on the next plane back to the States."

"It's a little overwhelming when you first get here."

"Talk shop for a minute," Ron said. "I hear you've got low breaks on the Training Facility footings, west end. Ha Li screwed up a big continuous placement."

"Breaks are fine," Mark said. "There was some question on the three-day breaks but the fourteen- and twenty-eight day breaks are fine. Had some pump problems on a big floor slab, but that's been ironed out, too."

"Good." Ron waved goodbye. "Got to get going. See you around." He made his way through the tables.

"Nice guy," Mark said. "Works in the office here in Jidda, but like all these office guys, they get way too much into our business if they smell problems. That's why the field guys never raise problems to them unless we absolutely have to. You wind up getting more help than you want."

Mark and Cathy got in the Blazer and made the long drive back to Tabuk.

Chapter 8

Fresh from his shower, Mark stepped out of his room into a velvety evening. Venus was brilliant in a vault of royal blue. He walked down the row of roaring window air conditioners to where the door to Dick's room stood open. The Wednesday night poker game was already underway.

"Come in, come in," Brian said.

Mark stepped into the frigid air and cigar smoke. Bill Vance, Kurt, Dick, Mike Robb and Jim Redding were seated at the dining room table. Poker chips, Saudi riyals, cards, ashtrays, glasses and bottles were strewn in a clutter in front of them.

"Join us, Mark, we need your weekly donation," Brian said.

"Hello, little buddy," Dick grunted. He sloshed more Johnny Walker into his glass.

Brian poured an inch of Johnny Walker Red into his glass. "Beer?"

"There are some bottles in a Sohat water bottle box in the kitchen," Dick said. "Help yourself, but first pony up some more money and join the winner's circle." He dealt the cards while Mark went to the kitchen.

"Sure." Mark slid an empty chair up to the table

between Jim and Bill Vance. "Cash some chips for me," he said, tossing three ten-riyal notes on the table.

"Not staying long?" Dick snorted. He turned his white smile and tequila-sunrise eyes on Mark.

Dick was playing a Rosemary Clooney tape – *shoo-fly pie and apple pan dowdy* – Mark stopped himself from making a joke about it.

Mark found the beer bottles in the kitchen. Beside it was another box, full of paperbacks. "Hey Dick, what are you going to do with this box of books?" Mark asked.

"I'm done with them. You remember Bill...Bill, what was his name? He worked for Lyon Associates."

"You mean Williamson?" Tom said.

"Not Tom Williamson, this guy was a real tall skinny guy. Wore cowboy boots all the time."

"Anyway," Mark interjected, "are you giving these books away? I see some science fiction titles in here I'd be interested in."

"Oh, yeah?" Redding folded his hand and joined Mark in the kitchen to peer into the box.

Jim and Mark pawed through the box. "I like these Heinleins and this Asimov," Mark told him. "You can have this Philip K. Dick stuff."

Redding fished out a handful of books and lit a Marlboro. "Yeah I like his stuff. He was a really paranoid guy, it's fascinating stuff. And he paints the picture of Northern California in the sixties beautifully. I lived in Sacramento then."

"Speaking of twisted reality," Dick called from

153

the other room. "Brian here has won five in a row. How twisted is that? Bring that other bottle of scotch out here would you little buddy?"

Mark stacked his paperbacks on the couch, handed Dick the bottle and sat back down at the poker table. Dick tossed five riyals into the pot. "And for another small fee, I'll show you my award-winning hand."

"How small a fee was that?"

"Five riyals."

Bill and Jim tossed chips into the pot. Kurt and Mark folded.

"No guts, no glory," Dick told them.

"I was working for Vinnell then, in Vietnam" Jim said. "Installing packaged power plants and water treatment systems out in the boonies."

"You get shot at a lot?" Mark asked.

"Not once. Safest job in the world as long as we never carried guns. The North Vietnamese knew they were going to win the war soon, and then they'd own this stuff we were installing. They protected us from the Viet Cong."

"Must have seemed boring when you went back to building shopping centers in Sacramento," Mark said. No wonder Jim doesn't care anything about reading adventure fiction – he's living it. "Of course, you could be back with your family."

Jim just studied his cards.

"Oh. Sorry," Mark said.

Dick nudged him, "Usually not a good idea to ask about somebody's wife back home. These things change quickly."

"Well I went back to Sacramento for a while, after

we lost the war, " Jim continued. "But the unions made things so bad I had to get out. I signed on with Intercontinent for a couple of years doing the highways in Iran."

"I was there, too. I liked Iran," Mike said. "With the Shah gone everything we built will start falling apart."

Mark pushed his thoughts of Jennifer out of his mind, folded his cards and poured himself home brew down to the sludge line.

"Mark! You playing or you going to sit there daydreaming?" Dick said.

"I think I'll call it a night." He shoved his chips into the pot. "Here's my weekly donation."

"I know where you're going," Dick leered, "Give that little blonde my regards."

Mark stepped out into the night with his precious books and walked to his room. He lay down on the couch and opened Asimov's *Foundation* and read for a while.

He began to daydream of being fifteen years old again, on a summer afternoon, the humid Midwestern breeze rustling oak and elm leaves outside. He lay on his bed reading and thinking he would travel the world one day. He began to feel restless, so he drove over to the HSC rec center. Through the glass wall of the snack bar he could see Cathy sitting with two other nurses.

She beamed at him as he walked in. He went to the serving line, got a can of Moussy and paid the Palestinian at the cash register a riyal.

Mark sat down with them, opened his beer and

took a sip, "Ah, the king of beers." He made a face to show it was a joke.

"Hello," Mark said to Elspeth and Moira. "Been doing any sightseeing recently? The gold souks in Jidda or Riyadh?"

"We haven't been off this bloody compound."

"Unaccompanied women aren't allowed to travel in this country," Cathy reminded him.

Mark squinted. "Good rule." He softened his joke with a smile, but the women did not return it.

"I bought some shrimp from our commissary," Cathy said shyly. "Are you hungry?"

He nodded slowly. "I could eat." Cathy stood up. "I'll go change clothes. Give me one minute." She scurried out.

"She's crazy about you, you know," Elspeth said. She turned her dark eyes on Mark. "All she talks about."

"Well, that's flattering, but..."

Moira and Elspeth exchanged looks.

"What brings you to the desert?" Mark said.

"Money. Unemployment is over twenty percent in Belfast."

"Belfast," Mark said.

"Bel-fast," Moira corrected. "Emphasis on the second syllable."

"Lots of good pubs on University Street," Elspeth said.

"I prefer East Bridge Street all the way from the opera house to Albert Street Bridge, marvelous restaurants, pubs, entertainments." countered Moira.

Cathy appeared with a huge net shopping bag

full of packaged food. For some reason that irritated Mark. The two Irish nurses stood.

"Well, we're off then," Elspeth said. "Pleasure talking to you."

They went out the double doors in their English-cut nurses' uniforms and practical British shoes. The half-dozen Jordanian and Palestinian men in the room watched them every step of the way.

Mark and Cathy went out and got in his Blazer.

"Arab men!" she snorted.

A wash of pink and gold remained in the western sky even though it was nearly nine o'clock.

At Mark's room, Cathy stocked the shelves and the refrigerator with the food she'd brought. "You know, you really don't need to buy so much food," Mark said.

She came out and sat down beside him, "Sorry. Just thought it would be more convenient."

"I eat over at the mess hall most weeknights."

She snuggled close and her full body kindled a liquid fire in Mark. After a while they moved to the cool bedroom and made love. The yellow light in the living room made a Matisse of the beige furniture and the blue carpet.

It was pre-dawn when Mark woke. Cathy was sleeping heavily beside him. The air conditioner hummed softly. Mark made a cup of instant coffee and went out to his Blazer. The air was skin temperature, absolutely still. The sky arced from black through pale blue to white in the east where the sun would soon rise. Mark sipped his coffee while sun came up and the air turned instantly hot. Then he got in the Blazer, drove to the Dormitory job site and walked through the second floor checking piping with the Korean quality control men. At nine he stopped and returned to his room where Cathy was lying on the bed reading a romance novel. "I've got things I need to do today," he told her. "We can get together tonight."

"Working?"

They walked out into the heat and got in his Blazer. He cranked up the engine and turned on the air conditioning full blast. "I promised Dale I'd drive to the beach with him, do some snorkeling."

She sat staring straight ahead. The Blazer idled, the silence stretched paper thin.

"Look, I know you're not crazy about Dale, but I like him," Mark said.

She put her hand on his arm, "It's alright," she said, "You go play stepfather to your overgrown son. I'll walk back over here later. Meet you at your room at dinner time. I'll make shrimp."

Mark knew he was trapped, "Sounds great."

He watched her go into the HSC compound gate. A month ago I wanted to spend every hour with her, now I feel like I don't have an hour to myself. He drove over to the little sheet metal building where two U.S. Air Force sergeants ran the U.S. post office for American government people stationed at Tabuk. There was a letter in his box. He got back in his Blazer and sat for a minute staring at the familiar small hand writing. He opened it and read the single hand written page slowly. Then he read it again right down to the last line '...think it best if we stop

158

seeing each other.' She had signed it 'Jennifer' not 'Love, Jennifer.' He folded the letter carefully and put it back in the envelope. He sat staring at the sheet metal building in the glaring sun remembering the years he and Jennifer had been together. All the way back to their first blind date when they'd both been undergraduates.

Then he drove to Dale's room at the HSC dorm.

Dale was standing out in front, like a kid waiting for the school bus. He was wearing cutoffs, huge white sneakers, tall white socks that came up to just below his knees, and a red Grand Island Racquetball Club tee shirt. He slung a huge Adidas gym bag into the back seat and climbed in smelling of scented soap.

"Ready for some snorkeling over at the Red Sea?" Mark said. "I brought the gear."

Dale's grin wavered under his mirror shades. "I'd rather go to Aqaba and rent some kick-ass jet skis."

"Sure," Mark said, sensing a Nebraskan's fear of deep water.

They drove out along the desert road to Aqaba. "By the way, Cathy and I would like to invite you over for dinner one night."

Dale watched the desert scrolling by.

To fill the silence Mark started talking. "This is the old spice route, camel caravans going to the trading centers in Damascus and Aleppo, or the Phoenician ports of Tyre and Sidon. In biblical times the Nabateans controlled both the East-West trade, silk coming from China, and the North-South trade, spices coming up from Yemen, where the queen of Sheba was selling frankincense and myrrh."

"Speaking of the queen of Sheba, it sounds like you two are spending a lot of time together."

"Yeah." Mark shot him a glance. "She's a nice person."

"Yeah," Dale said. "Well, sure, I'll come over for dinner some night. But I'd rather get off the compound, like this, or go out-of-country, than sit around talking with Cathy, or any of the nurses."

"You still got a girl back home?"

"Maybe."

They drove in silence for a time.

"We should go to Bangkok sometime," Dale said. "Every Brit at the hospital has been there, says it's paradise." He looked at Mark. "I'm only going to be here a year, I'm going to use my vacation week to go to Bangkok, let's go together."

Mark shrugged.

"Twenty-four hour party, man."

"I'm not sure I want a twenty-four hour party. I work seven days a week here most weeks. And driving to the Red Sea or Aqaba or Jidda is work too. If I'm going to take time off, I think I'd rather rest than party."

"Well, you can do that in Bangkok too. Lie around the pool and drink real beer, eat great food."

Mark thought about it for a while. "OK, let's go."

When they arrived in Aqaba, Mark parked in the Holiday Inn parking lot and started up to steps to the pool side bar, but Dale decided they needed to explore the town a little before they rented jet skis.

They set off together down the dusty street, moving uphill away from the Gulf.

"Surprised Cathy gives you any time off from playing house." Dale said.

"I got some time off for good behavior. You and everybody else seems to have a lot of interest in what we are doing."

"Cutest couple in town."

"Speaking of which, I want to get back to Tabuk in time for dinner."

Dale laughed. "Pretty short leash she's got you on."

The street was dusty and hot. Walled houses were set at random, streets were unpaved. They saw no one. A few hundred feet and they reached the main highway that came from the port, went up through the mountains and across the desert toward Amman and on to Damascus. It was clogged with a continuous chain of trucks grinding up the hill in low gear. Each truck towed a flatbed with a Russian tank on it.

"Let's cut across," Dale said. He dodged across the street looking like the former high school jock he was.

"Hauling Russian military supplies to Syria and Iraq," Mark gasped as they trotted up the hill. "Be another war one of these days."

"Hey, look. This is a park," Dale said. They walked through a dusty empty lot between threadbare mimosa trees. At the edge of the park was a low hand-built concrete wall. They sat down on the wall in the thin shade of the trees. The view was spectacular: the deep blue arrowhead of water between bare desert, a row of freighters laying at anchor, the town of Eilat backed by bare mountains. It was Friday, the Muslim Sunday, and very quiet away from the truck traffic; even the flies seemed less persistent than usual.

Dale pulled a slim joint out of his billfold and lit it. "This will improve the view."

"You shouldn't bring that stuff into Saudi. This isn't the States where the laws are a joke," Mark said. "You could get in real trouble over that stuff here."

They smoked the joint down to nothing, then wandered back down the hill past closed stores with dust-caked windows.

"Look at this," Dale squinted into a window. "A travel agency. Bunch of dumb Europeans pay money to vacation here." Mark squinted inside. Faded posters on the wall advertised the ancient city of Petra, crusader castles, the beach at Aqaba, the Intercontinental Hotel in Amman. "Those Scandinavian girls lying in the sun around the pool must create some real sexual tension in the Jordanians," Dale said.

"All of whom are supplementing their income by stealing things from the guests," Mark said. "Have you noticed the hotel windows won't lock? The front sliding panel will lock, which is all anybody checks. But these guys have twisted the inside of the back panel latch so it won't lock. They come and go through the window, helping themselves to your cash."

"That's something I like about Saudi, " Mark said. "It's safe to walk the streets there, and there's virtually no petty theft."

Mark and Dale changed into swimming suits in

Mark's Blazer then situated themselves on chaise lounges under umbrellas by the Holiday Inn pool. The breeze off the Gulf was cool. They drank an Amstel, then another one.

"You going in the pool?" Mark slipped off his tee shirt.

"Nah, I spent all last week in and out of our pool. When are you guys going to get that pump cavitation problem fixed? Every time we try to backwash the sand filters, the number two pump stalls."

"I designed the fix myself," Mark said. "Kendall, the contractor, refuses to do the work. If I were you, I'd have HSC's maintenance people do the work. I'll give you the design drawings. Now, no more shop talk, please. This is my day off."

Mark dove in, swam to the bottom and drifted up, the rippling rectangles of light pulsing silently around him.

Tonight, back in Tabuk, Cathy would be at his room making dinner when he walked in. That irritated him. If he could only get her to understand that he liked spending time by himself.

After a while Dale went down to the beach to rent a jet ski while Mark lay in the shade by the pool, daydreaming.

When they got back to Tabuk, Mark dropped Dale off and went back to his place to find Cathy laying out a big shrimp dinner on tiny plates. They ate, talked and made love, then he drove her back to the HSC dorm.

When he finally got back to his room, Mark felt like the day had been a work day.

Mark worked even longer days the next week since he planned to take a couple of days off and go with Floyd and Danielle on the trip to Damascus he'd agreed to.

Floyd picked Mark up at seven Thursday morning and pulled out onto the arrow-straight desert road to the border between Saudi Arabia and Syria. He kept the accelerator floored, the speedometer sat at 140 kph as he dodged nonchalantly around slow-moving trucks. Floyd chattered on about job site issues as he steered with one hand, throwing glances over his shoulder at Mark in the back seat. They left the mountains behind and crossed the featureless gravelly desert broken only by the occasional hulk of an abandoned Toyota pick-up truck.

"Shouldn't be more than two hours from the border to the hotel in Damascus depending on how many times I get lost," Floyd said. He slowed for a truck ahead, then eased over into the other lane like a soldier peeking out of a foxhole. Floyd floored it and they whipped around the truck and back into their lane. Floyd dodged around another truck without slowing. "We could stop and drink a beer at the rest house at the border if you like, real beer, not that home brew you guys make in garbage cans, but I'd rather press on."

"Yeah, let's press on," Mark agreed. Anything to get this driving over with.

Floyd swerved into the opposite lane to go around a string of eight trucks. In the distance, a car was approaching them. In typical desert driving style, both cars were going as fast as the engine would move them, combined approach speed of about two hundred miles per hour.

"Ease up, Floyd," Mark said uneasily. "I'd like to reach Damascus alive." Floyd dodged back into the right lane and the approaching car whizzed past.

Floyd snorted a laugh. "We'll get there."

"I'd like to have an epiphany on the road to Damascus, like Saul did. Think you can arrange that?"

"Epiphany?"

"Enlightenment. Of course, Paul was blind for three days and nights afterwards. Sounds like bad drugs to me."

"What the hell are you talking about?"

"Acts, Chapter Nine, I read it just before we came up here. I'm not a bible beater, but I do want to know about the history of a place before seeing it."

Mark chuckled at Floyd's uncomfortable expression. It's always good sport poking fun at these Christians, Mark said to himself.

They reached the hotel in Damascus, spent the night, and the next morning sat in a sunny courtyard eating breakfast – superb tea and Syrian bread with jam and butter from Denmark. Bees buzzed over hyacinth blossoms along the white rock wall. The staff was courteous.

"We're going to look at silver," Danielle said. "Silver is supposed to be cheap here in Damasacus."

"I'm going to find a 'street called Straight', the street the apostle Paul walked down." Mark said.

"Won't find no straight streets in this part of the world." Floyd buttered his bread with quick strokes, one on each side of the knife.

"That was the name of it in the Bible, book of Acts, Chapter Nine. God told some guy to go find Saul of Tarsus at Judas' house on a street called Straight. And he did. Restored his sight, too. I'm going to go find it."

"We're going to the silver souk, the Frezense store. We were told that was the best," Floyd said. At the front desk, he had the clerk write the name of the store in Arabic on a piece of paper. Outside, the doorman flagged a taxi over and they clambered into the ancient Mercedes. Floyd showed the driver the paper. The driver studied it for a long time. Then there was a conversation in Arabic between the doorman and the driver. "The Frezense store," Floyd repeated unhelpfully.

"Is here," the driver said.

"Let's go," Floyd said.

"Here," the driver repeated.

"He is saying it is nearby," Danielle added. They climbed out of the taxi and showed the paper to the doorman. He pointed down the street. "Fifty meters."

Mark decided to walk the other way. "I'll catch up with you at the store in half an hour."

The shops and the street retained hints of French colonial days. Through the glass windows of the Frezense store Mark could see Danielle seated at the counter, a cup of tea in front of her. Two impeccably dressed sales clerks were bending solicitously near. Floyd stood outside the shop, jingling the coins in his pocket. "Well, did you find your bippy?"

"No." said Mark. "But I did walk down the street called Straight."

Floyd gave him a puzzled look. "I thought you were kidding."

"I thought you were one of the little Christian circle, reading the bible and praying every Sunday night."

"I go because the boss goes," Floyd said. "Part of the game." He eyed the sky, looking for a way to change the subject. "People are real nice here."

"Yeah, they are. Shame the Israelis cause so much trouble. This whole area could be great."

"Shame these guys keeping starting wars they can't win," Floyd said. "Get beat every time despite all those tanks from the Russians."

"The U.S. supports and supplies the Israelis," Mark said. "Why we keep doing it is beyond my understanding. They do nothing but stir up trouble for us and for everybody else. We got sucked into supporting Israel when Britain pulled out in the late 1940s and left us holding the bag. Now AIPAC..."

"What – pack?"

"The Israeli lobby is powerful and effective. It's political death for any congressmen to oppose them, so the American taxpayers wind up supporting all kinds of aggressive Israeli activities, even when it's not in the best interests of our own country. It's ridiculous. One tactic they use is to equate lack of support of Israel with anti-Semitism, so if someone doesn't support the nation of Israel they are branded as a religious bigot, when in fact it's just the other way around." Mark shook his head and looked around the sunny street. "Don't get me started on American support of Israel." He looked Floyd in the eye, "Remember, I'm talking about the nation of Israel, not the Jewish religion, they are two distinctly different things, no matter how often the Israelis try to convince us otherwise."

Floyd stared at Mark. "That was quite a tirade."

Mark apologized, "Sorry, talking about religion is a bad idea."

"People have always fought about religion."

Mark cooled down. "You're right."

Danielle came out with her purchases in a beautiful box wrapped with red ribbon.

The shop owner followed her out, bowing. "Thank you very much. Have a pleasant stay in Damascus."

The next day they boarded a Syrian Arab Airways Ilyushin 86 airliner for the flight to Istanbul.

"Russian-made plane," Floyd said.

Mark watched Tarsus flow by below the wing, the blue of the eastern Aegean, and Ephesus, the lost city of Troy.

By comparison to Damascus, Istanbul felt and looked European. They checked into their hotel and Floyd announced, "We're going to the leather souk," he bustled Danielle into a taxi. "Meet us there. We'll go to dinner. I have the name of a great restaurant."

Mark sat in the dim lobby sipping strong Turkish coffee, cardamom and brown sugar in the sugar bowl. Outside, the street was deeply shaded with old trees. Around him men in dark suits without ties sat, coffee cups and newspapers at hand. Mark closed his eyes and listened to the language, felt the texture of the air. He was no longer in the Middle East. Outside on the sidewalk most of the men passing wore black leather jackets.

Mark took a taxi to the leather souk. The traffic was less frenetic than Saudi Arabia, but the cars were far older. He walked into the labyrinth of aromatic leather goods.

"Try it on, sir?" a man said from a booth. He held a brown leather jacket. There were mountains of jackets, handbags, boots, shoes, belts, everything that could possibly be made of leather in stall after stall after stall. Small boys and old men lounged on piles of new leather goods. Mark walked quickly to reduce the sales pitches.

He saw Floyd dithering around the front of a booth. Inside, Danielle was dickering over the price of another item. She had been there an hour and a half and had bought four items. The proprietor was a tough and experienced bargainer, but Danielle was giving as good as she got. Eventually they settled on a price of three hundred twenty five lira.

They departed with bundles wrapped in brown paper tied with string.

"I'm probably going to have to ask you to wear one of these coats back to the States next time you go, to avoid the import duty," Floyd said.

Mark crammed into the microscopic front seat of the taxi and piled the bundles of leather goods on his lap for the ride to the hotel.

That evening Floyd directed the taxi driver on a fruitless forty-five minute search for the restaurant he had heard about. As Danielle's comments became more and more pointed, Floyd's stream of jokes tapered off. Eventually they gave up and had dinner at a kebob place three blocks from the hotel – lamb and flat bread in heavy cream sauce. The waitresses were young and buxom, black-haired, beautiful, wearing heavy eye liner, brilliant smiles and distant attitudes. They all spoke English remarkably well. When he thought Danielle was not looking, Floyd eyed the waitresses and winked and bobbed at Mark.

The next day was windy and clear as they walked the wide plaza behind Santa Sophia with the Bosporus spread out before them.

"We don't need to go in," Floyd said. "I've seen enough mosques."

They had lunch at a tourist place overlooking the Bosporus, empty in the off-season, but the shrimp was good. They watched flights of pigeons wheeling past the famous blue-and-ocher domes and minarets. Mark refrained from talking about Constantinople, the Byzantine Empire, what the city might have been like when it was the center of the civilized world. German-made tour buses lined the street. Knots of tourists wandered across the expanse, lost in its vastness, as the emperor had intended.

Mark leaned on the stone wall and watched the choppy water under windy skies. Ferry boat horns bleated and the waves clacked and clattered on the stone wall. What if I just kept going, on and on, Mark thought, never going back. The sky was a deep blue laced with cirrus.

"We need to get down to Syrian Arab Airlines and get our tickets reconfirmed." Floyd pulled a sheaf of tickets out of his inner jacket pocket, then stuffed them back in. "The guy at the hotel said Syrian had a desk at the Sabena offices."

He bolted ahead of Mark and Danielle to get into the taxi queue.

At the airline office a patient clerk tried to explain. "You can change planes in Larnaca."

"But we would miss our connection. We must go to Damascus; the car is there."

"This flight, through Ankara," Danielle interrupted. "What's Hava Yolari?"

"It is the name of the Turkish airline," the clerk said. "Your ticket cannot be changed unless you pay the additional..."

"We would have to spend the night in Larnaca," Floyd interjected.

"The airline will pay," Danielle stated.

Mark grinned to himself. I wonder if Floyd will have this much fun when he's back home in Florida, dickering with a salesman over the price of a set of radial tires for that Cadillac he's so proud of. Mark wandered off to study the posters – the Red Sea resort hotels at Aqaba, the windmills overlooking the harbor at Mykonos. He had been to both places.

They had a two-hour flight back to Damascus, then into the car and across the desert toward the Saudi border under a wide orange sunset.

At the border, with its rows of trucks waiting clearance, Mark realized he didn't have his inoculation record with him.

He was taken to a room with a wooden desk and chair, lit with bare fluorescent tubes. An unshaven

Egyptian dressed in wool jacket and sweater, very slept-in, told him, "There is cholera in Tabuk."

The man's yellow eyes were cold. On his desk was a glass full of clear liquid with ten old-fashioned steel hypodermic needles. "You must be inoculated." He opened his hand at the needles in the glass.

Mark opened his passport and showed the various exit and entrance visas, "I work for the American government. We have had all inoculations."

Floyd bustled in with his inoculation booklet held open. "He's same as mine." He pointed to an entry. "Two months ago. Cholera, typhoid, tetanus – it's all here."

Floyd stood on one leg, then the other. He jingled the keys in his pocket, three clicks, then a pause, three clicks, then a pause. The doctor looked at one, then the other, then gave that characteristic nod of the head meaning "okay."

Outside in the cold night air, Mark breathed, "Thanks. You saved me from a dirty needle."

"We've got to get going," Floyd intoned, "Can't spend all night here waiting."

"I haven't had a cholera shot recently," Mark said suddenly nervous.

"Get your girlfriend over at HSC to give you one." Floyd eased the car up to the border guard and handed their passports to the Saudi kid in uniform.

"Cholera ain't nothing beside the health hazard from that rot-gut beer you boys brew." Floyd sat fidgeting until the guard dropped the passports back in Floyd's lap. The iron pipe barrier was raised, and they sped off into the night. Floyd kept the Chevy at a steady hundred and forty kilometers per hour.

It was midnight when they pulled up in front of Mark's room at the motel. He collected his suitcase from the trunk. "Well, thanks for inviting me along. I really enjoyed it."

"See you tomorrow," Floyd said and sped off.

His room felt hot and stuffy. Mark got the air conditioners running, opened a beer, and unpacked his suitcase. When he'd finished, he noticed a folded piece of paper on the coffee table with his name on it in Cathy's handwriting.

"Del tells me there's one of the married quarter houses available. She'll assign it to you if you ask."

Mark slumped down on the couch.

Chapter 9

It had been a fourteen hour day by the time Mark finished his last inspection and filled out his daily quality assurance report listing the piping and duct work accepted and ready to be covered with steel studs and sheetrock wall board. He drove home under a hazy indigo sky. The mess hall was closed, but he went to the back door and got the mess hall manager to have the cooks wrap him a couple of sandwiches to take home for dinner. He ate one of them driving to the little prefab steel post office.

There was a letter postmarked Columbia, Missouri – a letter from Allen and Carla Hayes. Mark drove to the house Del Winn had assigned him. He felt a great sense of relief when he saw the lights were not on. Cathy had said something about working the night shift this week, so he'd have a few nights to himself. He went in, turned on the lights and the air conditioner, hit the button on the hot water heater and took a quick shower while the house cooled down. Cathy had neatly arranged her comb and brush and shampoo in the bathroom. In the wardrobe in the bedroom she had left a pair of jeans and two tee shirts. She was keeping most of her things at her room in the HSC dorm, and would sleep there sometimes, but it was

174

clear she planned to spend a lot of time with Mark in this house. Mark stood looking at her clothes for a moment. What am I getting myself into? he thought. Then he dressed in jeans and tee shirt, padded out to the kitchen and cautiously removed a bottle of home brew from the refrigerator. He eased the snap ring back on the cork. It opened with a pop and foamed off a third of the volume into the sink. After the foaming was over, he poured himself a glass of murky beer and sat on the couch with his feet up on the coffee table.

In a scrawling hand, Carla described the winter weather: "…very rainy this fall, and now everything's frozen solid. We've had two ice storms already. Electricity out for a day both times." Mark grinned. She went on to say that prices for cattle were falling, but the price for hay was staying high. They asked when he was coming for a visit.

"Next May," Mark said to nobody. He wrote them a quick note. "I'll be back in Missouri next May and would love to come for a visit." He sealed up the letter and found a U.S. stamp for it. Then he refilled his glass with muddy home brew, leaned back, closed his eyes, thinking of how the Missouri countryside looked in the summer. The humid air stirring the elm and sycamore trees in the yellow afternoon sun, insects chirring, hay fields baking in the sun.

It will have been a year since I've been back there; I'll have a lot to tell them about. He drifted into sleep. He woke to a strange soft creaking from the roof of the house. The unfamiliar sound came again.

Mark dressed and prowled through the tiny house,

but there was nothing amiss. He peered out the window over the air conditioner but could only see the few square meters of empty backyard where Cathy was trying to get some grass growing in the sand.

He opened the front door. A gust of wind blew a haze of dust in. He slammed it shut and turned on a light. The room was already hazy with dust.

"Shemal coming," he muttered. These dust storms could last for one day, sometimes two. He got some towels and stuffed them along the top and bottom of the front and back doors.

By six the next morning the little house was creaking and groaning under a steady wind. Mark drove to the area office with his headlights on. Visibility was only about five meters in the blowing orange dust.

Vance was pacing the persian carpet in the waiting room. Danny, Dick, Mike Robb, Ed Preece, and Ray Barton, also a couple of Koreans from the Ha Li company were there.

"I don't want Ha Li setting steel in this wind," Vance said. "I'm going to direct a weather delay, but I want you to bring Mr. Kim in personally to meet with me. This morning. I don't want any misunderstanding about whether they can proceed or not like they did with night concrete last summer."

Ray Barton nodded. Ursula pulled a sheet from her typewriter and gave it to Vance. He read it, signed it, and handed it to Dick. Ursula snatched it back, stuck it on the copy machine and made four copies, then gave it back to Dick.

Redding stumped in, hard hat pulled low on his forehead. "Request a weather delay directive," he muttered.

Ursula was already typing.

"I'll have Kim in here at nine," Dick said on his way out.

"You want to see Mr. Kim?"

"Which one?"

"Dormitory."

"Too many Kims around here," Vance snorted at Redding. "Yeah, I want to see your Kim too. In my office this morning at ten. Bring along a list of tasks you think are critical and safe to continue in this storm and a copy of the network schedule. I'll have your weather delay letter ready when you get here."

"Can't keep the workers sitting around the barracks with nothing to do for too long," Dick said.

Redding nodded. "Yeah, pretty soon there are fights, drinking, vandalism, no telling what..."

"I know that," Vance said. "But I'm not going to get anybody killed or injured trying to work in this storm. It's supposed to clear in a couple of days." He speared Dick Davis and Jim Redding with a glare. "Both you guys got that? Your projects are shut down."

"Got it," Redding said.

Mark followed Redding's car, headlights on, driving ten kilometers per hour in the dust. At the job shack, Mark pulled a ghutrah off the back seat and wrapped it around his face and head leaving only his eyes exposed. Then he dashed though the dust to the shack. Every flat surface inside was already coated with dust.

"There's no sense you staying here," Redding told

him. "We're all on administrative leave until this storm clears." He glanced at the network schedule on the wall. "Go home. Take the schedule and the drawings with you and work out a way we can make up two lost days. You'll need to go over that with Mr. Pak tomorrow or the next day."

The house creaked and groaned in the sandstorm. Above the wall was only swirling beige-and-ochre dust. Mark closed the door and stuffed a couple more hand towels around the edges to seal it tight. He got a bottle of beer, a plastic glass, and spread the schedule and design drawings out on the coffee table, but his mind had drifted to autumn in the Midwest, imagining the dust storm outside was a cold autumn rain. He stretched out on the couch, remembering how the days turned cool, the leaves changed colors, MU Tiger football on the radio. In his mind's eye, he sat across a wooden kitchen table from Jennifer. They were in their own small house on their own farm in Missouri. The window behind her was open to the garden and the scents of spring drifted in with the breeze. She smiled, reached out and took his hand in hers. She had just come in from the garden; her long black hair was tied back, her white peasant blouse was stained with earth and green. "I love you, Mark. This is what I want."

A pounding on the door woke Mark. The house was still creaking and rattling in the wind. Outside a form huddled near his door in the blowing orange haze. It was Dick, his face wrapped in a red-and-white ghutrah. Mark pulled him in, slammed the door shut, and stuffed the towels back around it. "What are you doing out in this stuff?"

"Just trying to get out of the dust." Dick unwrapped his turban and made himself at home. The room was shadowy with the drapes pulled. The window air conditioner labored on.

Dick slurred, "You got anything to drink?" It was clear he was already drunk.

"It's ten in the morning, Dick."

"This storm calls for emergency measures."

Mark went to the kitchen and retrieved a bottle of Bushmill's scotch from under the kitchen sink, set it and a plastic Missouri Tigers cup on the end table. Dick shook himself as though he had come in from the rain, not the dust, then poured a generous portion into the plastic cup.

"Join me?"

"No thanks. Too strong for me."

Dick capped the bottle and paced the room. He lifted his glass, grinned his perfect grin. "Thanks, little buddy." Dick went into the kitchen, came back with a tall bottle of home brew and eased the cap off. He poured carefully into another Tigers cup and handed it to Mark.

Mark sipped "All right, I'll join you." He sipped again. "Wonderful stuff, high nutrient content." He pulled a dining table chair around to face Dick. "You don't have any real beer at your place do you?"

Dick laughed. "Heineken's gone the first week of the month. Hey, what happened to that little blonde over at the hospital, the one I introduced you to?"

The wind rattled the door and kept the drapes

twitching. There was a patina of dust on every flat surface.

"You guys sure keep track of my life, don't you? I'm going to go over to pick her up once she goes off shift." Mark improvised. He glanced at his watch. "About noon."

"Day like this is perfect for hanging out at the 'Down Under' in Ala Moana shopping center..." Dick said. "Great little bar. I know all the guys there."

Mark's mind wandered as Dick repeated one of his stories about the beach life in Honolulu.

Mark tried to call up his dream about him and Jennifer living the simple life they'd talked about so often, deep in the country.

"...happens next?" Dick was saying.

"What?"

"I said, when this job ends are you taking your girlfriend with you to the next job?"

"Well.." Mark said.

"Don't do it. Real bad idea. I tried it with my second wife." Dick was slurring his words. He stared at the coffee table. "Kiss of death for a relationship." Then his chin sank slowly down to his chest.

Mark looked at him sitting there. All those years of experience, Mark thought, all the people he's known, the places he's been, and this is where he ends up: drunk on somebody's couch in a construction camp out in the middle of nowhere.

Mark wiped dust off the table and spread the schedule out and worked for two hours on a method of making up lost time.

When he'd finished his schedule analysis, Mark

drove cautiously to the job site and discussed the plan with Redding,

"They may claim acceleration if we direct this," Redding said. "But if we can amend the structural steel change order to pay them to air freight the fitting in, we'd gain two days. And we'd only pay the cost difference on the air freight." By three in the afternoon, they'd worked out the plan. Mark got in his Blazer and eased cautiously around the site. The cranes and the open structure were ghostly in the howling dust. Then he drove to the HSC compound and rang for Cathy. She came down quickly, looking clean and fresh in a black-and-white knit dress.

She gave him a big smile. "Your timing is perfect," she said. "I just got off work."

At the house, Mark was pleased to see Dick had gone. He showered, and when he came out to the living room Cathy had dusted the tables and couch, had a frosted mug of beer on the coffee table and dinner started.

I should tell her now, Mark thought. Dale and I are leaving for Thailand in four weeks and I haven't even told her yet. But he had been procrastinating, knowing how hurt she would be. I don't want a full-time relationship, he told himself. Been there, done that. Cathy is great, Jennifer was great, but I just don't want a full-time girlfriend right now. But I can't seem to train myself not to acquire one.

"You seem pensive," she said over dinner. She looked beautiful in her black and white dress.

"Just thinking. This kind of lifestyle, moving from place to place. Intentionally severing yourself from friends, and moving on to a new circumstance, new friends. And then doing it again in a year or so. It's the nature of construction overseas."

"Overseas health care is the same," she said.

"Is it good for us?" Mark asked.

She smiled, pressed herself against him, *J'aime* perfume-scented.

"I wonder if we don't pay a price we may not even recognize," he said.

"We do it by choice. There's nobody forcing us to do this."

"Yeah." He waved his beer mug around vaguely. "But sometimes I wonder…what's the alternative? Stay put our whole lives? Stay in the States, in the same job, in the same town? That's no good. But after you've been overseas for a while, you can't go back to where you were and start right in like nothing has changed. You're different. That's why a lot of these guys just keep moving forward. Doesn't really matter where, just keep moving. Each country is different, but all construction camps are basically the same."

She studied him, unsure of where he was going with the conversation. "It's the same in nursing."

Mark continued on, "I'll bet everybody in this camp has told each other about their hometowns back in the States, about how they are going to go back there someday, retire, settle down. Things will be just as they were when they left. But they won't. And things can't ever be just the way they were."

"Sounds like you're trying to convince yourself of something," she said, avoiding his eye.

He stepped forward and hugged her, "I'm just philosophizing."

After a minute, she hugged back. The pink on her neck faded out of her pale complexion.

"Too many questions," she said. "But there's more to life than playing poker, drinking, running off to Thailand with the guys."

Good, she knows about Thailand, Mark thought, refilling his glass, hesitated, then changed tack, "Well, being married over here doesn't seem to work too well either. Look at Tom and Trish Farris."

"That was his own fault. He's so pathetic."

"Kind of ironic. The one guy who wants a long-term relationship doesn't have one."

"He's going to have to make some changes in himself if he ever hopes to have one," she said.. "Nobody was surprised Trish left him. He's so selfish, so full of himself, so artificial. I don't like him at all."

"Speaking of not liking someone, why does your friend Mary dislike me?"

Cathy gave Mark an appraising look. "She thinks you will hurt me."

Mark looked at the cloudy beer in his glass and held back a sigh. "It's not my intention. How about a peace offering from me to her. The three of us go for a weekend in Aqaba. Or a day trip out into the desert."

Cathy gave him a kiss. "Great idea. It's not that she disapproves, she just worries."

He put his glass of beer down and put his arms around her and pulled her close. Later with the Roberta Flack cassette playing, they slid into the cool bed and made love. Afterward he said, "Let's avoid these painful conversations, okay? I don't want to lose today by always talking about tomorrow."

"You can't always live just for the moment."

He hugged her. "We're back into the conversation I want to avoid. Can we stop talking?"

They lay still. Her breathing was just slowing into a soft snore when there was a pounding on the door. Cathy snapped upright in bed pulling the sheet around her. "What is it?"

Mark pulled on his jeans and went to the door. It was Redding, ghutrah wrapped around his mouth and nose to keep the dust out. His car was idling, lights glowing, in the whirling dust. "Been an accident, we need to get to the job site."

Mark told Cathy to wait for him, dressed and drove at five miles per hour through the streaming dust.

At the site, there was a cluster of vehicles, their lights aimed at three toppled precast concrete panels. Mark wrapped a ghutrah around his face, clapped on his hard hat, and made his way to the huddle of people. Mr. Kim the general manager was there along with four of his senior staff. One of the twenty ton mobile cranes came lumbering up through the dust, two Koreans walked in front of it as ground guides. An HSC ambulance arrived, and one of the Lebanese doctors came over. "Two men were trying to tighten the props on these four panels. Wind blew them over. One man is still under there," Redding explained.

A Korean expertly fixed the rigging on the crane to the concrete panel's lifting lugs. The drum started turning slowly and the panel rose, skittered sideways. "Get more guidelines on that thing," Redding snapped, and the Koreans scrambled to clip two more guidelines to the edge lugs. The crane eased the panel up. The wind swung it around and more men scrambled to add their weight to the upwind guideline. The crane maneuvered the panel to one side, wood chocks were laid out to support it, and it was gently set down. Then the other fallen panel was lifted and stacked on top of the first. In the dirt was what looked like a pile of pale blue clothing, a pale blue Dae Joon uniform, dark along one side. Mark and Jim Redding stood aside as the doctor and two Koreans examined the twisted form, but it was clear there was nothing that could be done. A plastic tarp was brought, two workmen wrapped the body in it, several men lifted it into the back of the ambulance. "Bring to the hospital," the doctor told Mr. Kim and Jim Redding. "The police will come."

"That means we'll all be held at the hospital until their reports are done," Redding said. The doctor tilted his head in the way that meant yes. Redding turned to Mr. Kim, "You should go back to your quarters, Mr. Kim, so the police don't take you to the police station. You need to be here at the jobsite tomorrow." Redding pointed at Mr. Pak, the architectural quality control man. "Mr. Pak and I will go with the police and stay with them until they've completed their investigation. You get out of here too, Mark. Notify Vance and start on the fatality report. And get back here to the jobsite tomorrow morning to make sure they don't try to keep working in this wind. You're in charge until I get back. Keep Cooley and the rest of them from blowing this thing out of proportion. Next thing you know we'll have an IIAA investigating team up here from Jidda and we'll lose even more time."

Cathy was waiting up for him when Mark came back to the house and he told her what had happened. "I've got to go let our people know what's going on, probably be gone at least an hour. Try to get some sleep."

Mark went to Vance's house, roused him and told him the story. The he drove Vance to the office. They tried to radio the Jidda office but reception was so bad they couldn't get through. "You go get some sleep, then get back to your job site. Tell Mr. Kim I want him in my office with you or Redding at nine o'clock," Vance said.

Mark drove home, took a quick shower to get the dust off and slid into bed. Cathy was a warm presence beside him. He lay still for a long time, but sleep still eluded him.

"Thinking about the man who got killed?" she whispered.

Mark sighed, "Yeah. How it can come so fast. One minute you're in the middle of your life and the next minute its over and everything you've planned doesn't mean a thing. All gone."

Cathy stroked his head.

"I think about how that poor guy looked," Mark continued. "Crushed to death. When the crane lifted that panel off him, he didn't even really look much like a person anymore, just a bloody mess in his uniform." Mark's stomach rose and he fought it down. "Death isn't noble, it's just ugly and dirty."

"I know," she said softly.

After a while he was able to get to sleep.

When he woke, the storm was beginning to subside.

The entire day was spent on the accident report. Redding appeared at noon and went back to Vance's office with Mr. Kim. By evening the air had cleared. Dae Joon had front-end loaders working to clear the drifted sand. "Can start work?" Mr. Pak asked.

"When Mr. Kim comes back with Mr. Redding," Mark told him.

It was almost five o'clock when Redding got back. "Start work tomorrow morning," he said to Mark and to Mr. Pak. Mark stumbled out to his Blazer and drove to his house.

Cathy had made spaghetti. After they ate they sat finishing off Mark's only bottle of real red wine in the light of the mismatched candles.

"Yesterday and today haven't been much fun, but in general I find I really enjoy this work. It's very satisfying. The design drawings are beautiful, and bringing those designs into physical reality is really enjoyable. Projects have a beginning and an end, and at the end of a job you can point to what you have accomplished. That's missing in so many other kinds of work." He felt a feathery brush of irritation when she agreed with him so quickly.

"On the other hand, these construction guys are

nice guys, but not real deep, you know? I know that sounds very superior, and I don't want it to, but except for the work, I don't have a lot in common with them. Sometimes when we're all having breakfast together up at the cafeteria, I look around at these coarse, friendly guys and wonder how I got here." He finished his wine.

"I'm glad you came here," Cathy said very softly.

Mark turned off the lights and they sat together on the couch. "Of all the guys I've met here, the one I respect the most is Jim Redding. He spent three years in Vietnam working for Vinnell building power plants. Little diesel stations to power individual villages. Said it was the safest job in-country. By then the Vietnamese knew they were going to win the war and they wanted the power stations intact. As long as he never carried a gun he was perfectly safe. Only time he was in danger was pay day when he drove around from site to site in a jeep with a footlocker full of cash and two armed guards. That's where he met his wife, Ti, up near Da Nang."

"I thought he told me he was married, back in Sacramento. I think he said he had two daughters."

"I guess that was before he went overseas," Mark said. Cathy looked away. "Don't know. He worked for a while for Intercontinent doing airfield upgrades all over Micronesia. Truk and Peleau and Guam and such."

The flame from the two candles stood almost unmoving.

Mark knew he was talking too much but couldn't

seem to stop. "All these guys, racing off to work for construction companies in all these remote sites all over the world. Sort of a brotherhood. After some years they all kind of know each other, moving around from project to project, company to company." He paused. "And the construction companies themselves? They're phantoms. They don't own their own offices or their own equipment or have their own employees. They hire people for a 'job', then let them go. Lease all the office space and equipment, then let it go at the end of the project. All these guys move from job to job, never going back to the States, never working very long for the same company. They don't fit in to stateside society any more.

Cathy watched Mark's eyes in the yellow candle glow. "We're all just ghosts," he whispered.

Chapter 10

Mark slid out of bed and turned the air conditioning up a notch. Five a.m. and the outside temperature was already over a hundred degrees according to the thermometer he had stuck up outside the window.

Cathy stirred under the sheets. "You're not going to the job already, are you?"

Mark hesitated. "Yeah, just for an hour or so."

"I thought you said you had today off?"

"I need to be available whenever the Koreans are working, and they only take one day off every two weeks."

He took a quick shower in tepid water.

She had a cup of tea ready for him when he was showered and dressed.

"I should be back around lunchtime. Let's go up to the mess hall for lunch." He didn't wait for an answer, but took his tea out to the Blazer and drove to the job site. Redding was in the temporary office, booted feet up on his desk, hard hat on his head, Marlboro in his mouth. Mark decided to launch right into his sales pitch.

"Look, Jim, this new fuel facility is a big change to the basic contract."

"Supplemental agreement," Redding corrected.

"Right. So it's important it be managed carefully. I'd like to treat it as a separate subproject under the Dorm building contract."

"No. I want it fully integrated into the Dorm building quality control operation we've got set up and finally running smoothly."

Mark set his hard hat on the desk and leaned across Jim's desk, dodging Marlboro smoke.

"Like you said, it's running smoothly. The Koreans are going to treat the fuel facility like a separate project. They are going to assign Mr. Ah as Quality Control chief, Mr. Pak is going to be foreman. I want to be the project engineer."

"I need you here."

"I will be here. But I'll also be overseeing the fuel facility. The mechanical inspections for the Dorm are routine, Kim knows what to do and he's doing a good job of it. All I have to do is show up for the final testing of each subsystem. The next big push won't start until air conditioning installation and commissioning and that's months away."

"They given you a schedule for the fuel facility yet?"

Mark laid his trump card on the desk in front of Redding. "Here's a network diagram of the whole thing based on the Koreans' block diagram schedule. If they get NTP by the 20th, welder certification should be done by the 25th, welding and erection for one hundred twenty days, then testing for ten days concurrent with pipe testing, final system test May 10, then ten days of painting and finish site grading. Everything done by May 20th."

Redding eyed the diagram. He pulled the Dorm project master schedule out of his desk drawer and studied it in silence.

After a while, he stubbed out his cigarette. "Okay, you've got it. Give me a QA plan, schedule of three-phase inspections, draft QC org chart, submittal register, and draft payment schedule by Wednesday."

Mark grinned and clapped his hard hat back on. Exhilarated, he left the office and got to work.

The afternoon flew by. That evening he whipped the Blazer up to the front of the house and slammed inside. Cathy was in the kitchen boiling shrimp. He grabbed her, lifted her off her feet and gave her a kiss. Then he put her back down, ran his hands over her breasts and gave her another hug.

"Must be having a good day," she said brightly.

"Pour us some beer and I'll tell you about it."

Mark stripped off his sweaty clothes, and took a one-minute shower, all the hot water the tiny flash-heater could provide, dried himself off and dressed in clean jeans and tee shirt.

Cathy had a couple of frosted mugs full of home brew on the coffee table when he came back to the living room.

"I've been assigned project engineer on a two-million dollar fuel facility. Site work, piping, tank welding, testing, pump installation, electrical, finish work, test and acceptance, everything." He raised his glass in a toast.

After a while they adjourned to the bedroom and made love. The cool blue of the drapes masked the orange light of dusk outside. The window air conditioner hummed.

Later, Mark lay staring at the sagging ceiling fiber boards while Cathy dozed. He put the job site things he'd been thinking about out of his mind and concentrated on listening to his heartbeat. He lay perfectly calm. Satisfied. Maybe this is true happiness, he thought, or is there something more?

Two weeks later the Koreans had steel plates piled on the site of Mark's project. A grader was cutting grades while a water truck and two one-ton rollers were compacting the fill where the fuel tanks would be sited.

Mark noted with satisfaction that the lifts were a uniform twenty centimeters, and the color of the sand indicated the moisture content was about right. He went into the temporary steel building Dae Joon had put up for an office. All four Koreans saluted him as he entered and he returned their salutes. They had a desk set up for him in the corner near a roaring window air conditioner. There was a bottle of Evian water on his desk. He placed his hard hat next to the water.

Mark spent fifteen minutes with Mr. Pak, drawing a draft schedule of the project in network form. "Now redraw this to scale and post it on the wall. Update it each Saturday."

He drew another sketch with the purchasing and testing milestones inserted. "Also show each submittal form 2544 with fifteen days for review and approval, then do a backward pass to adjust all the start dates."

Mark put his hard hat on. He went out into the fiery heat and walked the entire site. Mr. Ah hurried over, saluting as he ran. "Good morning, sir."

"Morning, Mr. Ah. Is the QC plan finished?"

"Yes. In office." They went back to the office and Mark took his time going through the plan. Three Koreans had worked all night writing the plan. The English was stilted but understandable and very clearly lettered. Mark added four notations for omitted items. He got his copy of the job site safety manual out and tore out the four pages in the welding safety section.

"Good job, Mr. Ah. Please make these changes and add the words on the pages of this safety manual. Then one copy for you and one copy for me."

Mark got in his Blazer and drove off feeling satisfied, then pulled over. He'd forgotten welder qualification.

He found Mr. Ah under the sunshade at the welding area. "When is welder qualification?"

"Quantification?" Mr. Ah looked puzzled.

"Welder testing."

"Testing," Ah said. "Yes."

"When?"

"Job site."

Mark spread the drawings and specs out on the table and read from the specifications: "Each welder must be qualified." He pointed to the words.

Ah read them slowly. "Just a minute." He went out into the heat and came back in a minute with Mr. Pak.

Pak removed his dirty Mickey Mouse gloves and they exchanged salutes. Pak and Ah talked for a time in Korean.

"Six welders," Pak said to Mark.

"We need to have weld testing done here at the job site," Mark said.

Pak and Ah talked this over. Mark took the opportunity to get out his weld testing handbook. He showed them the illustrations of ductility test coupons before and after welding and bending.

Mr. Pak and Mr. Ah studied the diagrams and nodded.

Mark left the site and drove back to the house, dusty and sweating. Inside, the window ACs were running, but it was still too warm. The place smelled bad.

"Is that you, hon?" Cathy called from around the corner in the kitchen. The house was hot and close and full of the smell of boiling shrimp.

"I can tell we're having shrimp for dinner." He started for the bedroom to change.

"Sorry." She came out drying her hands. Her smile came and went quickly.

He smiled reassuringly, touched her shoulder and retreated to the bedroom where he put on sandals and a clean tee shirt.

She had a frosty mug of home brew on the coffee table when he sat down.

Why should it bother him if she called him 'honey'? His own irritation irritated him.

They ate at the coffee table in the living room with *"An Officer and a Gentleman"* on the Betamax. The shrimp was tender and flavorful, and Mark told her so. The Rimalaud sauce was tart and cold.

The coffee table was so low they had to hunch forward to eat, which Mark disliked. And he didn't like the way Cathy put so much food on each of their plates, small plates, salad plate size, which she heaped with food. He'd mentioned at least three times that he preferred full-sized plates with small amounts of food on them. "It looks better. And we can always go back for seconds."

But she didn't change. He stared at his full plate, wondering why he was whining. After a moment, he put a smile on his face and made conversation.

After dinner, they walked arm in arm along the row of tamarisks to the rec center and watched the evening movie, a nondescript and unfunny comedy – after twenty minutes they snuck out giggling. Walking home was pleasantly cool. There was even a breeze.

"I hear it might rain sometime soon," Cathy said, pressing close to him. "Winter is here."

At the house, he put Roberta Flack's *Blue Lights* cassette in the player and they went into the silent bedroom and made love with just the light from the kitchen seeping into the darkness.

"I love you," she whispered.

He hated the lilt at the end of the sentence; the expectant pause, as she waited for him to say the same thing. He let the silence answer.

Why don't I make some commitment to her? His life felt very full as he lay there in the darkness, a sheet drawn over their bodies. She was a welcome warmth in the darkness beside him, breathing softly, and that's all he wanted, in his heart of hearts, just the moment, nothing more. All the rest seemed too complicated.

After a while, he eased out of bed and turned the cassette player off. He sat on the couch in the darkness for a time listening to the hum of the air conditioner.

Wednesday night the poker game was in Dick Davis' room.

"Raise one."

"Fold."

"See you and raise one more," Brian said.

Mark tossed a five in the pot and pulled three riyals out. "I'll pay to see what you've got."

Kurt glanced around the table. "Two pair." He laid his cards out neatly, the pair of fours to the right, the tens to the left. Dick, Bill and Mark tossed their hands into the discard pile.

"Flying high tonight, Kurt," Jim said. He collected the cards and started shuffling.

"I can't believe you raised with less than two pair," Dick said to Brian.

"Calculated risk."

"Speaking of calculation..." Dick held up his wrist. "My new watch. It's got a calculator built right into it. Take a look. I got it down in Jidda last week."

"How much was it?"

"About three hundred riyals."

Kurt's eyebrows rose, his big square-toothed smile widened.

Dick lurched up to pour himself some more Johnny Walker. He shook Kurt's shoulder, "It's only money. You shouldn't care, all the money you're taking for us negotiating these change orders."

Kurt's smile disappeared. Money was a serious topic with him. "IIAA owes our company lots of money for all the change orders you people have directed. I'm negotiating them all out. But I don't get paid on commission, just salary."

"Good negotiator is worth his weight in gold to any construction company," Redding said, dealing. "Kendall management back in the States knows that. You sure they don't pay you a commission on all the money you collect for them?"

"No," Kurt snapped. "But my job is to collect every penny IIAA owes Kendall legally and fairly." He poured himself a generous shot of Johnny Walker from the bottle Dick had set on the table. "That's the only reason I stay out here, working fourteen-hour days and living like this. And putting up with you people." He softened it with a smile.

Dick nudged him, "Come on, little buddy. You don't put up with us, you love us."

Bill carefully fanned his cards one by one in his hand. "Speaking of putting up, would you guys put up or shut up. The game is seven card stud."

"Boring," Brian interjected.

"Nothing wild," Bill continued. He tossed a riyal out. "One riyal ante. None of this Baseball, Dr. Pepper, Old Maid, Go Fish, or whatever other crap it is you guys come up with. Let's play some good honest poker."

They all anted up.

"Raise you two," Dick said immediately.

"You haven't even looked at your cards yet." Bill gave Dick a raised eyebrow glare.

"Hey, you play your hand and I'll play mine," Dick said. "Two riyal bet. Let's see your money. Hey, Mark, you still seeing that little blonde from HSC?"

"Yeah," Mark said. He folded his cards.

"You going on leave next week?" Brian asked.

"Yeah."

"You got all the below-grade concrete work done at the fuel facility?" Bill asked, directing his attention equally at Jim and Mark.

"Ninety percent," Jim said promptly. "Still two thrust blocks to be placed at the west end. We're waiting for the okay on the soil stability from Geotech. We may have to do a half meter or so in over-excavation." Jim sucked his Marlboro. "At the Dorm building, we've got the forms and falsework done for the elevated slabs on the entire second floor. Going to try a two hundred cubic meter continuous placement."

Vance eyed him.

"That's too much," Brian said. "Ha Li tried it two weeks ago, in that parking garage floor slab. Couldn't do it. They were still placing concrete at three in the morning. The batch plant was screaming about the overtime their crew was running up. They'd been on the job eighteen hours by then. Mr. Kim finally came out himself and told them to put in a construction joint and stop for the night."

"Yeah," Brian added, "I was out there that night. One of their pumps broke down, then a truck got stuck on the ramp down to the placement, then their other pump broke down. It was a circus."

The Johnny Walker bottle made the rounds; glasses were topped up.

"Always have a back-up pump."

"How many pumps does Ha Li have?" Jim asked.

"Three," Dick said. "But one's a dog. They've got two good ones. You could probably borrow..."

"Rent."

"Right. Rent one of them for Sunday night. Have your guy talk to Kim Dae Soo; he's the senior QC guy."

"And now I hear Ha Li's got low breaks on that slab concrete," Brian said.

Jeez, Mark thought, these office guys are constantly finding fault with the field work. They should try working in the field themselves sometime, with guys like them looking over your shoulder. Brian shuffled and dealt.

Redding shrugged. "I was over at the lab the other day. Their fourteen-day breaks are good; it's just curing slow, that's all. They'll probably have 3000 psi at twenty days. They just didn't get the three-day strength."

"Oh?" This was clearly news to Vance, "Well, maybe we should direct they not remove falsework for a week."

"You'd get a delay claim, even from the Koreans," Dick said. "Forms and scaffolding are on the critical path."

"So, uh, where you going on leave, Mark?" Brian's tinted lenses masked his eyes.

"Thailand."

A leer came across Brian's face. He nodded knowingly at the rest of them.

Wish this guy would mind his own business, Mark thought.

"Thought you had domestic responsibilities these days."

"Yeah, well, maybe I need a break from them, too. And I definitely need a break from you guys." Mark finished the hand and collected his riyals, "Got to go."

Mark drove to the HSC compound, picked up Cathy, and drove her back to his house.

Thursday morning was hot and bright. After breakfast Mark and Cathy found themselves sitting in the living room wondering what to do with their day.

"You know, we ought to go over to Dick's, have a few drinks, then go to the pool before it gets too hot."

Mark handed Cathy a beer and sat down beside her. "This couch is never going to get any softer."

"Do we need to go to Dick's this early?" She hid her face in her beer mug.

"It'll be fun. I think he starts every weekend day with a bloody mary and works from there. Did I ever tell you about the time he and I were making the rounds of the bars in Aqaba and he left a twenty dinar tip for the bartender? I think it was at the upstairs bar at the hotel next to the Holiday Inn, the Meridian."

"He's going to end up old and broke."

Mark put his beer down and pulled her to him. Her kiss was warm and wet. He had never liked its wetness.

She sipped from her beer again. "If we're going to the pool, I'm going to wear my swimming suit under my clothes." She went into the bedroom to change.

Mark peered out the window between the sheers and venetian blinds. Over the yard wall, the sky was white and hot and dusty. No clouds, never any clouds.

He leaned back and closed his eyes, remembering Missouri and how the air had smelled when summer thunderstorms rolled in from the West. Birds would dart from tree to tree as the afternoon breeze died before the storm. Then a cool gust of air mixed with the smell of hot pavement and the rumble of thunder.

"I really don't feel like seeing Dick and all those Kendall company guys this early."

Mark came back to the present, opening his eyes reluctantly. Cathy stood in the bedroom doorway. He hated the way she stood with her too-wide hips and turned-out toes. In a minute, he thought, she'll offer to cook something just so we don't have to see other people.

"We could go to the pool and come back for an early lunch," she said with false eagerness.

I hate that tone, he thought, and her hairstyle, she needs to do something with her hair.

Mark downed the rest of his home brew and put the bottle back in the refrigerator. "Look, I'm going to run over there for an hour or so," he said. "Maybe

Dick's got some Heineken he'd trade us for a bottle of Scotch."

She didn't say anything.

"I'll be back in an hour and we'll go to the pool for a while before lunch." He turned and walked through the door, feeling her silence boring into his back as he left.

The next Friday morning, Mark left Cathy sleeping heavily in the bedroom, fixed himself some coffee and stood at the window admiring the patch of grass that had started growing in the tiny walled back yard. It took an hour of watering every day to keep it alive, but when watered, the sand was surprisingly fertile. He stepped outside, but the flies soon drove him back inside.

A piercing European car horn blared.

"What the hell..." Mark muttered, imagining one of the Koreans injured in a job site accident. Cathy stuck her head out of the bedroom, "What is it?"

Mark went out the front gate and there was an old white Mercedes 230. Tom Farris sat in the driver's seat, under an open sunroof, shirt off, grinning his big-toothed grin. "Bought this car the other day. Runs great. Let's go to the beach."

"Now?"

"Yeah," Farris said. He ejected the Neil Diamond tape. "I've got food and sodas in the cooler, It's a clear bright Friday, our day of rest, let's go."

Mark shrugged, "I don't know, man..."

"What's to know? Just do it. And if you don't like Neil Diamond, I've got lots more tapes." He scrabbled around in a pile of cassettes on the back seat.

"Cathy's here."

Tom gave him a grin over sad eyes. "Hell, bring her along, man." That was a quite a concession, Mark knew. Tom didn't like Cathy, and she didn't like him.

Mark knelt down at the car window. "It sounds like fun, man, it really does, but I need some advance warning."

"No you don't. Be spontaneous, the open road, the sun, the beach, no worries..."

"Not today, Tom. I've got to go to the job site."

"Friday?"

"Everyday."

Tom shrugged and pushed the Neil Diamond cassette back in the player, "By the way, I'm going with you guys to Bangkok. Dale invited me. I need a vacation."

Mark slumped. "Yeah. Well, have fun at the beach." Tom put the car in gear and drove off.

Inside, Cathy busied herself in the kitchen, "I hate that man. Those sad puppy-dog eyes over that big fake grin."

"He's lonely." Mark said. "His wife left him. I need to go to the job site today."

"I don't know why you always seem to have to go to the job site," she said. "And why you have to go to Bangkok with the guys, and why you don't..." she started crying, covered her face, and ran into the bedroom.

Mark drove to the job site.

The Koreans had fabricated a shed out of steel and packing crate wood. In the shade, they had the shop drawings laid open and fifteen men were laying out the steel plates, marking cut lines with chalk, and cutting lines with three slicing saws. The noise was deafening. Other men were grinding neat bevels on the cut surfaces using heavy side-grinders. With a five ton crane and come-alongs, men maneuvered the steel plates on to wood blocks where the welding crew members were welding plates together into subassemblies. One man was welding on temporary lifting lugs for hoisting the plates into place with the cranes.

Mark nodded to Mr. Pak and gave him the thumbs-up. Other companies cut steel plates with oxy-acetylene torches – faster, but it heat-treated the steel and created areas with a high probability of corrosion and cracking.

Mark and Mr. Pak had agreed on each of the weld designs. Pak handed Mark a welding mask and he watched a pipe flange being welded onto an elbow. The fire of the molten metal neatly followed the beveled valley of the two edges. When the weld was complete, the welder chipped off the slag revealing neat silvery fish scale weld loops built up just the right amount over the steel. After it had cooled, he'd grind it smooth and paint it with orange primer to prevent rust.

"Very nice," Mark said. Mr. Pak said nothing. Mr. Ah smiled and said, "Thank you."

The screech of grinders, the sputtering of welders, and the smell of the desert blowing in on the hot wind all created a clean, orderly sensation.

The work was highly organized and moving fast.

Two small mobile cranes were handling the steel sheet and fittings, laying them out in assembly order. All the steel was too hot to touch in the glaring sun. The workers wore leather welder's gloves, Pak and Ah wore the ubiquitous white cotton gloves.

Mark checked the shop drawings, consulted his Hobart handbook on the welds, looked at the amperage settings on the Lincoln welding machines, and checked the low-hydrogen rod in the drying ovens. He checked the lenses in the masks to assure they were dark enough for the welds being made. Then he checked the tank assembly yard where two Liebherr mobile cranes were laying out the steel plates that would form the fuel storage tanks.

"Use tag lines with the cranes," Mark told Ah. He flipped open his safety manual and showed him. The desert wind could move the steel while the crane had it suspended, crushing a hand or cutting off fingers.

Mr. Ah made a note and discussed it briefly with Mr. Pak. Mr. Ah had a degree from National University in Seoul and his English was much better than Mr. Pak's, but Mark knew Mr. Pak was the organizer and director of the entire operation.

Except for raccoon-eyes from sunglasses, the skin on Mr. Pak's face and hands was burned toast-brown by being outside all day, every day. He was the sort of take-charge, hard-working, no-nonsense, get-it-done kind of guy that existed in all construction companies, in all nationalities, that made it all work. His soft-spoken, terse authority was unquestioned. The men followed his direction without question because it was clear he had done all this before and could analyze and organize the work to minimize wasted time and effort.

Mark took a welding mask from Ah and watched a man welding a pipe connection. Once the pipe was set, welding had to be done from all angles – above, below and around. At first there was only blackness, then yellow lights appeared as the welder struck an arc. The fireball slowly travelled down the neat valley of the two beveled edges of steel, filling them with glowing weld metal. The arc moved around the joint, down the vertical face, along the bottom, up to other side, then snapped off.

They all removed their masks. The workman took a pointed hammer and chipped the slag off. The weld was beautiful – perfectly even laps of cooling weld metal, stronger and more ductile than the base metal itself.

"Very well done," Mark said.

Men were setting the third panel of the tank sidewall using a crane and come-alongs to spring it into place, then tack weld hand-size squares of metal to hold it in place while the finish welds were completed.

"If more than a three-meter span, use a spreader with the crane," Mark reminded Ah. He got a blank look. Back at the worktable, Mark opened the crane book and showed him.

"Tomorrow let's check dimensions, horizontal and vertical," Mark said. Pointing to the design drawings and using pantomime, he showed Ah what they would do tomorrow.

The day was cooling and the wind slowing as Mark drove back to his house. He'd spent all afternoon at the site. If they keep up this pace, I may have it done a day or two ahead of schedule. And I want it accident-free.

But the Koreans tend to conceal accidents, so I need to assure all safety regulations are followed, but if I keep harping on it, and an accident occurs, they'll hide it from me.

He pulled up in front of the house and went in.

As he went in to take a shower he noticed Cathy's comb and brush and shampoo weren't there. If she moves back to the dorm, that might be best, he thought.

Another cold, cloudy day dawned – winter in the desert. In the morning there was a fine misty rain for an hour, then silence. No wind, just overcast and cold. There is nothing gloomier than the desert in the rain, Mark thought as he hurriedly packed for Thailand.

Cathy was working the night shift for four days, so Mark was spared saying goodbye to her. He drove over to the HSC compound to pick up Tom Farris. Dale had taken yesterday's flight, so Tom was riding with Mark to the airport in Amman.

Mark was glad Cathy wasn't around. He'd told her he'd be gone for eight days, and the silence that followed was not comfortable. She'd decided to work some night shifts the day after his announcement.

Farris was standing in front of his dorm room with

his suitcase. He threw the bag in the back seat and plopped down in the front. Mark started down the straight highway to Amman in intermittent rain, eager to get away from the compound, the job, Cathy, everything. As they drove, Tom couldn't stand the silence so began to talk. When Mark only answered his questions with a grunt, Tom shifted to talking about himself and then about his marriage. He admitted Trish wasn't coming back. She had sent him the divorce papers just last week. For the better part of an hour Mark endured the story of the decline of Tom's marriage.

Tom finally wound down and put a Dave Mason cassette in the player.

"Jesus, change tapes," Mark said. "All that whining about losing the love of your life. We're supposed to be on vacation, not crying in our beer about past girlfriends."

"Speaking of beer," Tom said, "I'll be ready for one at the border rest house."

Mark shook his head. "I don't want to sit in that bare truck stop with all the other Turkish truck drivers drinking bad beer out of dirty glasses. Let's press on to Amman."

They got to Amman at dusk, found the Palace Hotel, and checked into a cold room with two narrow single beds. Their flight was not until the next morning at eight. The hotel bar was closed for some undefined reason, so they wandered down the street in a cold drizzle in search of a drink. At a dirty store jumbled with boxes of Lebanese foodstuffs, cheap clothing from China and other miscellany, they bought a pint bottle of something that turned out to be vodka. Standing in front of the store out of the rain, there appeared to be no alternative but to go back to their dingy room.

"Hey, here's a movie theater," Tom said, pointing. "Let's watch a movie."

The feature film was just starting. It was a cheap Korean-made superhero film subtitled in Arabic, with production values that made B-movies seem lavish. The theater was cold and almost empty. Mark and Tom sat there passing the bottle back and forth, staring at the screen.

After a while Mark whispered to Tom, "I'm going back to the hotel, get some sleep."

They made their way back to the hotel and lay on the single beds, each lost in their own thoughts.

Tom's the saddest man I've ever met, Mark thought. I wish he hadn't invited himself along.

Chapter 11

As soon as the plane turned off the runway at Bangkok's airport, passengers got up from their seats and began wrestling bundles out of overhead bins. Only Europeans and Americans follow rules, get in line, take turns. Mark's shoulders and neck ached from sitting on the plane, he wanted to brush his teeth and wash his hair. He looked at the deep green of the trees beyond the rusted hanger roofs and smiled. The window fogged over slowly, whiting out the concrete and the rusty sheet metal roofs and the white clouds in a Florida-blue sky.

He jammed his Thai International airline ticket in his paperback and made his way down the aisle littered with abandoned newspapers, plastic cups, tissue, blankets, and pillows.

Outside, walking across rain-cracked concrete to the terminal, the sun felt friendly, not the flat deadly white glare of the desert.

Inside, the boots of hard-eyed Thai customs officers in tight brown uniforms clicked slowly up and down the line at immigration. The terminal was high ceilinged, echoing, cool. He claimed his battered green Samsonite suitcase and pushed through the chattering crowd of families to the white picket fence that kept greeters back from the baggage claim carousels.

"Tour, sir, floating market, downtown, girls. This way. Best price."

Mark pushed past tour guides and into the first taxi in line. Despite the prospective tour guide hovering outside, Mark kept the door open.

"How much to the Windsor Hotel?" He asked the driver.

"Sixty-five baht, sir, good price."

Mark closed the door on the tour guides. "Okay, let's go."

The white Toyota moved down the ramp to the highway.

Thailand, Mark thought, where everything is available. And I have the money and time to enjoy it. No limits. The bright Cezanne landscape rolled by – rusty sheet metal roofs and muddy water, a crowded bus laced with Thai script, rain clouds in the distance, surreal in the brilliant sunlight.

Jet lagged, Mark rolled his window down to get some wind on his face...

"No air condition, sir?"

Mark rolled his window back up. "Air condition. Yes."

He dozed in the taxi, dreaming of an autumn weekend he'd spent with Jennifer. They had been loaned a friend's farmhouse in northwestern Missouri. Autumn was almost over, the colored leaves turned to brown, blown down by cold wind and rain. Both Jennifer and Mark felt something was already changing in them. Graduation would come next year, and their two years together, students at the University, would end.

They drove up to the house and let themselves in. It was small and well-lived in but uncluttered, with a fireplace, a rocking chair, no TV. They had walked up the hill behind the house, through the oak trees to watch the sunset on the flat horizon, then returned to the house and ate dinner at the old wooden table in the light of a kerosene lamp.

Mark wanted to say thank you, to Jennifer for giving him so much, but he didn't have the words. He touched her hand on the polished wood of the table. He wanted these moments to go on and on forever, and knew they could not. The glow in her eyes told him that she felt it too, that the time they had shared, the mundane things they had done together, would be among their best memories.

They built a fire in the fireplace, wrapped themselves in the quilt from the bed. Later they opened the double door to the bedroom and let the fireplace warm the room before they crawled into bed and a dreamless sleep. Far away in the night, a coyote yipped and howled once.

I wish I could stop time, Mark thought, keep these perfect timeless moments endless.

Mark started awake as the taxi turned down a side street off Sukhumvit Road.

"Where are you going?"

"Too much traffic."

They pulled up under a building's arcade and

Mark heard Creedence "*Green River*" playing. A line of taxis sat in the shade, their drivers sitting on the curb smoking and talking.

Mark tapped the driver on the shoulder. "No stopping. Go to the hotel."

"Stop here first. Girls, beer, smoke marijuana."

"No. Windsor Hotel."

Dale was waiting for him in the lobby.

Away from the rattle and roar of two-cycle engine traffic on Sukhumvit Road, the day was hot and still. Brilliant white clouds were building over the rooftops. The afternoon was drenched in sunlight and humidity. Mark wiped his face on his Levis polo shirt and slipped his dark glasses back on as they walked down the battered sidewalk.

"Where are we going?" Mark asked.

"Next door. The bar in the Federal hotel is better."

"Here we are." Dale turned into the driveway of the Federal Hotel. A couple of Toyota taxis were parked in the shade of the front entrance. Window air conditioners roared, condensation running across the sidewalk in a rust-stained trickle.

"Taxi?"

"No, thanks," Mark said. They pushed through the glass doors into the lobby and went into the bar. It was cool and quiet. American, European, and Japanese tourists, the Western men with their wives, the Japanese men in groups, sat at bamboo tables. The bar was decorated in tropic-modern with lots of large philodendron and banana trees in rattan planters.

Mark and Dale took a table near the window overlooking the swimming pool. A stunningly beautiful Thai girl, long black hair, slim body in white dress, wonderful long lashes over delicate Asian eyes, came to their table. She laid two Thai International Airlines coasters on the glass tabletop.

"Do you have Kloster's beer?" Mark asked.

"Singha beer," she said, her accent thick and beautiful. It didn't seem to be a question.

"Anything but Singha," Mark said. "Amarit beer?"

"Yes."

"Pepsi for me," Dale said.

"One Amarit, one Pepsi." She went gracefully to the bar.

"You know what I'd like to have right now?" Dale said.

"Yeah, I do," Mark said.

Dale flashed his tongue-between-teeth smile. "Nope, even more than that. A big glass of ice water."

"That'll probably have to wait until the next time you're back in the States."

"Or maybe Europe," Dale said. "We're still going skiing in Switzerland this winter, right?"

"Sure. But I don't think you'll find ice water in Switzerland, either. Nobody but Americans have ever heard of putting ice in drinking water."

Their drinks came. Mark took a big swig. "Tastes good," he said. "What is it about this place? I feel both full and hungry all the time."

Dale laid the straw aside and downed half his Pepsi with a big farm-belt flourish.

"Coming out of the desert, it's sort of like going from sensory deprivation to sensory overload, you know." Mark slugged down some more beer. "Back in Tabuk, it's a major event to get any real beer at all. That home brew is dog-piss. And the food...."

"...and women?"

"Well, yours aren't bad over at HSC. But over at the IIAA compound...nothing."

Dale laughed. "Things can't be that bad. You look pretty happy playing house with Cathy."

Mark rubbed his eyes. "Yeah, well look, we're on holiday so let's don't tell war stories from back home, okay? How about another beer?" Mark caught the waitress's eye.

"Why not?"

"Two more Amarits."

"Let's talk about what we're going to do."

Mark smiled. "Hell, I've only been here about two hours and I love this place already."

"Yeah, me, too, but I've got another idea. You're really going to love this."

Their beers came.

"See that travel agency down the hallway there? Well, I was in there yesterday. They've got round-trip airfare to Katmandu for 24,000 baht. Less than $240." Dale said.

Mark finished one bottle of Amarit, set his glass aside and started on the next. Thailand. No sooner have you gorged on one treat than you started on the next.

"Hey, I've just got to Thailand, and you want me to go to Katmandu right away?"

216

Dale smiled his teeth-and-tongue smile again. "Let's just walk down there and find out what they've got, okay?"

"After we finish these drinks, man. We're on vacation. We've got to remember not to hurry."

"Yeah. Pour some of that beer for me, would you?" Dale held out his empty glass.

"And there is one other thing." Mark said.

"Yeah, what's that?"

"The main reason we came to Thailand. You remember? Women?"

"Let's go talk to the travel agency first, then we'll hit the bars."

Dale wandered off to thumb through the postcards on the rack in the gift shop down the corridor. Was there a girl waiting for him back in Nebraska? For all his party-animal jock-talk, he seemed very cautious about diving into the fleshpots of Asia. But he's alright, better than Dick or Ray or the Brits back in Tabuk. The Brits have a kind of coldness about them. Their drinking seems more...closed, colder. Chasing hookers is a regular part of their life, not an exotic treat like it is for us. Maybe it's because they are older, been overseas longer, and the years have cost them their connection with their homeland. Like Redding said, they can't go back any more. They don't fit in. Mark wiped the frown off his face. I'm overseas, enjoy it. I don't want to do this forever. Soon I'll go back to the States, get married, settle down, buy a house.

Dale was waving from the corridor. Mark paid for the drinks and they went into the tiny travel agency office and sat on red plastic chairs while another stunningly beautiful Thai woman behind the desk studied her computer.

"Leaving tomorrow, eight o'clock, Nepali Airways flight 2751 arriving ten o'clock local time," she said. "Return Tuesday 1 p.m. Arriving Bangkok 7 p.m."

Mark pulled his American Express Gold Card out of his billfold, and they closed the deal.

Outside in the white glare and humidity, Mark flagged a taxi over. "Now let's go downtown."

They had the taxi drop them at the first bar they saw on Pat Pong Road. "This looks good. The Whiskey Au Go Go." They paid the taxi and walked in.

Baby Love was blasting out of the speakers above the red-lit door.

The place was packed with Western men. There were nude dancers on platforms at both ends of the bar. Five topless Thai girls sat on bars tools among the men. Topless waitresses in gold sequined bikini bottoms circulated through the crowd.

A topless waitress with a tray had Dale by the arm and was pulling him toward a table where three Thai women were sitting.

Dale broke free and made his way to the bar. Mark followed.

Another sixties song came on the stereo. "Did you see the size of the tits that waitress had? Thank god for silicone!" Dale shouted over the music.

Dale spotted the American beers on the bar. "Two Budweisers."

"Too bad they don't have Coors."

Mark felt a warm pressure on his side.

Her face was only fair, but her body was beautiful in its sequined bikini. "Hello, Sailor. You want company?"

Mark burst out laughing – a line from old movies. He put his arm around her and gave her a hug.

Dale nudged him. "I can't believe you. You picked the ugliest one in the place."

"I didn't want her to feel rejected; besides, I'm only interested in her for her mind."

More beers came. Paul Revere and the Raiders' *Kicks* played in the background.

Mark grabbed both her breasts, "Hey, Dale, look at this," he said, holding her warm breasts in his hands.

She smiled all the time, but her black eyes said nothing. She exchanged a few short words with the bartender, and led Mark upstairs.

Twenty minutes later, Mark stumbled out of the hotel into the darkness. There was a line of taxis across the street. He dodged through the traffic.

"Taxi?"

He climbed in. "Windsor Hotel."

"You want girl?"

"No. Hotel."

The taxi maneuvered through heavy traffic. Even at – Mark studied his watch in the darkness – 2 a.m. The city gliding by was a black-and-white diorama. Thai men sat in wooden chairs in bare rooms under bare light bulbs playing dice and cards. Fluorescent lights in stores crowded with jumbled Chinese merchandise, the colors all bleached to black and gray and white.

The glare from the lights of a filthy street side automotive repair shop lit the taxi for a moment. Mark noticed that his face reflected in the taxi window was not smiling.

A deep relationship or total freedom. Were they mutually exclusive? Is loneliness the price of freedom?

I am trying to do both by compartmentalizing my life geographically – exchange freedom in the short run for loneliness in the long run. But if I had brought Jennifer along, I think we would have been lonely together in our house at Tabuk. The distance between us would not have healed by moving to a new location. I would have been envious of the single guys, flying to Thailand like this, having any woman I wanted, lots of them. Mark remembered the look both Brian and Floyd had given him when he mentioned he was going to Thailand.

The cab whipped into the Windsor Hotel entrance. Mark clambered out, handed the driver two 50-Baht notes and went inside.

He silently unlocked the room door. Dale sat up in his bed. "Hey, man, don't forget we're leaving at seven in the morning."

"No sweat." Mark set his watch alarm for 6 a.m., undressed and slid into the other bed.

His sleep was heavy and filled with dreams. He and Jennifer sat on a picnic blanket under an oak tree overlooking Hulen Lake. The afternoon and had been warm and sunny, but now storm clouds were rolling up over the hills behind them and the lake water had turned a dark grey.

"I guess we should go," he said, and began gathering up the picnic things, but she just sat there staring at the water.

He tried to pull her to her feet. The wind was strong now in the trees above them. She turned and ran into the woods. The picnic things rolled and fluttered in the wind.

He struggled through the woods. Wind and rain came down, branches lashed at him. He could see no sign of her blue sweater and brown slacks.

He struggled on for hours as the day turned to evening. Eventually he went back to the car, and just sat inside, staring at the gathering darkness. The storm had passed and now there was only a dark, dripping silence.

Then she was there beside him in the car. "It's just for a year," he said. "But the job offer is not for family or dependents. I have to go alone. Just for a year."

"Why do you have to go at all, Mark?"

He searched for an answer and could find none. If he loved her, as he said he had all these years, he should at least be able to explain to her what he was doing and how their lives would fit together. But there was no answer, only the dripping of rain in the darkness.

Mark's watch alarm chirped in the darkness and he pried himself out of bed. Dale was already showering.

They left their suitcases with the desk clerk. "We'll be back in three days."

As they paid the bill, Dale asked, "Should we wake Farris and tell him we're gone?"

"Let's leave him a note." Mark pulled out a pen and wrote on hotel stationery: Gone to Katmandu. See you Friday at the Orchard Inn in Pattaya. He signed it, handed it to the desk clerk and climbed into a white Toyota taxi. "To the airport." Dawn made the tropic sky gaudy with orange and red.

Traffic was a clogged mass where Sukhumvit turned to Ploenchit Road.

"What's all this traffic?" Dale asked.

"Work day," the driver said.

Mark rolled up the window to escape the diesel fumes.

The traffic eased a little after they turned onto Ratcha Parob Road.

Dale had a joint rolled and stuck in his mouth.

"You look like some old-time cowboy out of a Sergio Leone movie," Mark said.

Dale tapped the driver on the shoulder. "Mind if we smoke?"

"Okay."

"Like we really need to get high at seven in the morning," Mark said.

"Best thing for us."

Mark saw the driver watching them in the rearview mirror as they puffed and passed the joint. Pretty soon the smoke cloud in the car was so thick Mark rolled the window back down.

They were moving along better now as they

reached the outskirts of Bangkok.

"You going to the States?" the driver asked.

"No. To Katmandu, Nepal."

"Ah."

The taxi was out of traffic now, rolling through lush green rice fields. The clouds seemed impossibly white in a blue sky. Mark shivered with the sensory pleasure of the colors and the cool air as the cannabis settled into his blood.

"Good dope. Thai stick?"

"Yeah," Dale said. "Some Brits had some down at the uh…what's the name of that disco in the Miami Hotel…anyway…" He paused to take a huge hit off the last of the joint, then passed it back to Mark, who took a minimal hit and tossed it out the window.

"We're going up to Nepal to hunt tigers," Dale told the driver.

The driver glanced at them in the rearview mirror. "Many tigers in Thailand."

"Not in Bangkok," Dale said. "Only pussies." He and Mark roared with laughter.

"Not in Bangkok. Tigers in villages."

Mark noticed the taxi was only going 40 kph in the fast lane of the highway to the airport. A steady stream of traffic roared around them on the right, horns blaring. The driver ignored them.

"When night is coming. No light…" the driver continued.

"No electricity?"

"Yes." Trucks roared past.

Dale turned and studied the traffic. "Okay, now go."

The driver jerked the Toyota over into the slow lane. A truck dodged around them, horn blaring.

"Tiger comes at night," the driver continued. "In the trees." The driver's hand perched on the rearview mirror imitating the prowling tiger.

Mark was laughing so hard he couldn't breathe.

"One man walking." The driver had both hands off the wheel showing the tiger's claws.

Mark elbowed Dale. "The driver's as high as we are."

"Coming one man," the driver continued with his story. "The tiger...down" One hand jumped from the rearview mirror to the back of his own neck, fingers curved into tiger's claws.

"Gang!" he shouted. His tongue stuck out in the victims' death agony. Mark and Dale were hysterical with laughter. The taxi pulled up at the departure entrance of the airport.

Chapter 12

The Nepal Airways 727 came down the valley between mountains and touched down on the single
runway. They rolled past a couple of dusty hangars
and a control tower, then turned around and taxied
back to the terminal. There was no taxiway. Feels like
Ozark Airlines coming into the old airport at Columbia when I was an undergraduate, Mark thought. The
distant mountains were no longer visible through the
haze that hung over the valley. The airport was on a
mesa in the valley. The city of Katmandu lay below,
beyond terraced rice fields.

"*Namaste*," Dale said, putting the *Nepal Travel-
ler* complimentary magazine into his carry-on bag.
"Welcome to Tribhuvan Airport."

They took a taxi into town and found a room at
the Eden Hotel. In their dingy room Dale pushed a
smoldering hash pipe at Mark. They stood passing
the pipe back and forth until the hash was gone. "Got
to get some more from room service," Dale joked.
The gray room and the thin dusty Nepalese air had
taken on a warm glow. Outside their room, in the
alley behind the hotel, somebody hawked and spat.
Dale and Mark exploded with laughter. "Nepalese
national anthem," Mark gasped out between gales of

laughter.

Eyes still streaming, they made their way down-stairs and out into the still, dry sunshine of afternoon. A Nepali dressed in a soiled brown suit materialized behind them. "Tour, sirs? Best prices, temples, moun-tains, villages. See Everest."

"No, thanks," Mark said, pushing past him.

"What we need," Dale said. "is a valley girl to go with this valley town."

"I'd settle for some lunch first."

"Well, this map lists an 'American Cafe' down this path."

A tiny woman in Nepali sari stepped around them. The top of her head was no higher than Dale's belly. She had an infant in an orange-and-red cloth carrier, a silver necklace and nose clip. She smiled as she passed.

"Hello," Mark smiled back at her. "People sure are friendly here. Not like the Middle East."

Dale pointed toward a cylindrical tower in the hazy distance. "That must be Bhimsen Stupa, so Parker Square is this way." He started off in the di-rection the tiny woman had gone, a giant in a land of tiny Nepalis. A child wearing only a sweater, no pants or shoes, stared at them.

"Hey, look at this," Mark said. He had unfolded his tourist map from the hotel. "Even this tourist map lists 'Freak Street'. Man, the sixties are still alive..."

"Party!" Dale whooped. "Everybody get high."

Mark folded the map and stuck it in the hip pocket of his Levis. "I was thinking more...spiritual. Pure. This town really was everybody's destination." He

touched the pale yellow plaster of a shuttered building wall. "I wonder how many of those old hippies
made it."

"Hey, look at this." Dale pointed to a faded poster
in the front glass of a hole-in-the-wall shop next to a
tee shirt shop. "White water rafting. We ought to try
it."

Mark and Dale wandered into the tee shirt shop
where cheap white Chinese tee shirts had been embroidered with Buddhist temples, or the word 'Katmandu' under a fan of marijuana leaves, or the Buddhist eye, or the tree of life.

"Best quality, sir. May I show you this also?" The
tiny tailor ushered Mark into the back room where a
younger man worked an ancient foot treadle Singer.
The whole shop was about the size of a closet.

"Custom embroidery." He laid out obviously
American tee shirts and jackets, one after another
across a stack of cardboard boxes like the flapping
of a wing. The clothes were worn, the embroidery
beautiful.

"How much?"

"For these patterns, fifty rupees. For another pattern..." He leaned his head to one side and opened
his hands.

"Let me look at the other tee shirts again."

The air in the shop was still and hot and very
close.

"How much for this one?" Mark asked, holding out a bright orange shirt with Buddhist eyes in
red and blue on the back. He wished the little tailor
wouldn't stand so close.

227

"Twenty rupees."

"I'll give you ten rupees for this one. And you embroider the tree of life and 'Kathmandu' on this one." Mark touched the sleeve of the Dos Equis beer tee shirt he was wearing.

The little tailor dropped deep into thought. "No, sir, twenty for each."

Mark put the shirt back on the rack and turned.

"But I can let you have this one for fifteen…"

They closed the deal at twenty and fifteen. Mark stripped off his shirt and put the new one on. Dale was outside waiting.

"There's supposed to be some nice temples back down those back alleys," Dale said. "Besides, the European women are probably staying at those youth hostels we just passed."

"But first we eat."

They had breakfast for lunch at the American Cafe – scrambled eggs, toast, and sausage – under calendar photos of the New York skyline and the Golden Gate Bridge and the "Please Don't Smoke Hash" signs. Dale talked about his party days at the University of Nebraska, his Blazer with BBS wheels, racquetball tournaments.

"I notice you never talk about your girlfriend," Mark said, and immediately wished he hadn't.

Dale frowned. "I'm not sure I have one anymore." Outside in the pale air, they started off at random. "You know, another thing I want to do is get the back of my Levis jacket embroidered."

So they had to go back to the hotel and get the jacket and then back down the alley where the tee shirt shops were.

"I like a lot of these designs, but these cheap Chinese tee shirts look like shit."

Finally, after Dale and the tailor had reached agreement on a price, standing out in the alley since Dale was too big to fit in the shop, they started off.

"It feels good just walking down some back street in a foreign city. Not going anywhere particular and plenty of time to get there," Mark said as they walked. They were trying to keep the pace slow to keep from getting winded in the thin air and also aware of how ludicrous it looked to be hurrying to nowhere. Mark thought they looked ridiculous anyway, Dale standing head and shoulders taller than anyone else on the street. His bare thighs below his cutoffs were the diameter of the tiny Nepali women's waists.

"You know, sometimes I think about just disappearing," Mark said. "Vanishing. Don't tell anybody, just walk off down some humid Bangkok street and take up a new life somewhere."

"You've talked about that before," Dale said. "You really going to do it? What are you going to work at? Open another American Cafe?"

"Nah. I'd roam the world for a while, then sneak back into the States, go back to Missouri. But I'd be someone else then, a new person."

"What's the point?"

"I don't know. Sometimes it just feels right." Mark fell silent as they trudged down narrow alleys and through sunshine-filled markets.

As they walked, the realization came to Mark, I've already gone to a distant country and become a different person.

After a while they stopped and sat against a temple wall in the thin sunshine.

"Hard to believe we're in Katmandu, isn't it?" Dale said.

"I always knew I'd come here. Back in the sixties I thought about it a lot."

"I didn't," Dale said. "In North Platte there were a couple of kids with long hair, but they were kind of weird anyway. We only thought about Nepal as a hash source."

"Certain places have a certain aura," Mark said. "And this is one of them. The stillness of the air, the color of sunset, the mountains beyond the rim of this valley. It kind of draws me here. Spirit of place."

"Well...," Dale said, slowly leaning against the ancient brick, his eyes closed. "That's all bullshit. I don't believe in any of that crap, karma, all that."

"Just good old Midwestern Methodist crap, right?"

It was silent, no sounds of people, traffic, no birds. The air was still.

"Back in my senior year at the University of Missouri," Mark said, "I shared a trailer with a couple of other guys, and Sunday evenings we'd sit around with our girlfriends, drinking beer and listening to the stereo or watching the TV with the sound turned off. One night the Sunday night movie was a film called *Sands of the Kalahari* with Stuart Whitman. Something about a plane that crashes in the desert in southwest Africa. We started talking about what it would be like alone out in the desert. One of my roommates was a long-haired guy, well into flower-power, the be-ins, the music, the psilocybin, the mescaline, the acid. He said he would do it. Take a year to himself for spiritual cleansing and meditation. I wonder if he ever did."

"Those were the good old days," Dale said, missing the point.

"Anyway, that conversation crossed my mind this morning. I'm in Tabuk for a year, spending my year alone in the desert."

"Not quite alone." Dale gave Mark a look, then slipped his aviator shades back on.

They wandered down back alleys, past crowding houses and empty lots littered with trash.

Bhimsol Tower stood ocher above the mists of morning. Mark slowed, stopped at a dusty, nameless street corner, the lush mystery of the world rising up in his imagination the way it had when he'd been a kid reading Robert E. Howard's *Conan*.

Mark and Dale stopped in a tiny tour office and bought an overnight hiking tour package for that afternoon. "Bus takes you to hiking path, you hike to cottage, three kilometers, and spend the night. Bus picks you up next morning, brings you back to Katmandu city square," they were told.

They walked along the path on the side of a mountain, going slowly in the thin air. The sunshine was warm and the air completely still. The people they passed all smiled and put their hands together saying *Namaste* in greeting. Mark felt a wonderful serenity, but Dale was tiring. All his gym-trained fitness seemed to have been lost somewhere. But at least he isn't talking, Mark thought.

They reached the cottage as the sun touched the mountain tops. The cottage was a hand-built mud brick building, one small dining room furnished with unpainted wood furniture, six tiny sleeping rooms each with a bed and washbowl and a plastic bottle of water. A sign told them where the outhouse was. There were no towels, soap or toilet paper, but there was a threadbare sheet on the bed and two wool blankets.

"Dinner. Yeah, dinner," Dale said. He frowned at the lumpy plaster on the wall of his room.

"You don't seem very enthusiastic. Not for somebody who's mentioned food at least five times since we left Katmandu," Mark said. He lifted the dirty brown curtain over the window. Through fly-specked glass, the valley was a hazy monotone. The light was fading fast with the sun behind the mountains.

Dale lay down on his bed, his huge tennis shoes hanging over one end. "Just knock on the door when you're ready to go to dinner."

"What did the guy say? Six o'clock?"

Dale gave a minimal shrug, his eyes on the ceiling.

Mark went out into the still dusk, down the promenade. There were voices from another room. Other than that, nothing moved except a bird gliding deep in the sky, far out over the valley. In just ten minutes it had become dark.

Inside, he switched on the only light, a dim bulb in a dusty glass box in the middle of the ceiling. He looked at the bathroom, at the thin blanket on the bed, the tile floor. How cold would it get tonight?

Two women were already seated at a long wooden table set with cheap Chinese plates and silverware. They were conversing in German.

"Hello," Mark said.

"Hello. How do you do?" the stocky one with short blonde hair said. She was wearing a short-sleeved white blouse with the kind of angle-cut sleeve Mark disliked.

A tiny wrinkled Nepali in dirty yellow clothes began serving a meal of thin lamb stew over rice. The rice was excellent.

"Have you been in Nepal long?" Mark asked. Aside from the small sounds of the tea-boy in the other room, it was absolutely silent. The room was bare except for table and chairs, two candles lit the room yellow. There were no windows.

"Just a week," the woman said. They introduced themselves. Mark stared at the candles, he felt fully and completely at peace.

"...went skiing, but could not come back," the dark-haired woman said with a laugh.

Dale leaned forward over his unfinished rice and vegetables. In the candlelight his smile was huge.

"Would you like to smoke some hash?"

The women exchanged a long look. Silence filled up the cold room, the wind whispered outside, the Nepali houseboy clicked cooking ware in the kitchen.

"That might be very nice," the woman with dark hair cut in bangs said finally. She smiled a brilliant smile. The stocky blonde kept her eyes on her plate, or on her friend, or on the smoky wooden rafters. She muttered something in German. Her friend shrugged.

Dale filled the hash pipe and passed it around.

"It's okay. I mean, I find I like the desert. I went overseas for the travel, not for the money..." Mark stopped talking to the blonde – she said her name was Urse – who whispered something to Matte, her dark-haired friend.

"Here," Dale tried to pass Mark the pipe again.

"That's enough. I've had enough," Mark said, but took a hit anyway. He watched the yellow candle flame stand, and shiver, and lengthen, and ripple, finding an ever-changing allegory in the silent flame. The silence deepened in the cold room. Mark imagined how it would feel to be warm in Urse's bed in her room while the moonlight stood clear and cold and blue on the snow. But how to make the first move? How to reach out and touch her?

"It's getting colder," Dale said to no one in particular. Urse hugged herself, breasts moving deliciously under her white cotton blouse. She started to slip her sweater on, then stopped.

There had been a sound in the silent room. Not distant. A rustling here in the room...

"Did you hear something?" Dale said. Mark's heart was pounding. A wild animal? Driven by hunger down from the snowfields? Urse looked alarmed,

Matte curious.

Dale went to the door, "Nothing moving outside."

He was halfway back when it came again. It was near.

"Under the table?" Mark said. Their chairs slid back in unison. The shadows were confusing.

Matte let out a whoop. "*Der hund!*"

Mark looked at a friendly dog peering over the table from the empty chair at the foot of the table.

Mark woke to cold silence. His watch told him it was 5 a.m. He dressed quietly, wrapped himself in a blanket from his bed and walked up the path behind the cottage. A hundred meters along the path, he stopped on a crest and watched the first touch of sunlight to flash the mountain tops to glaring white above a narrow band of gold. He stood transfixed as the line of light moved almost imperceptibly down the mountain snow in absolute quiet.

The clarity of the vista was staggering. Mark stood for minutes, ignoring the cold. He thought about how big the world was and how varied and strange and wonderful. He thought about all the things he could do with his life and all the things he had done. He thought about how lucky he was to be alive.

After a while he started back down the path to the cottage. He found himself first mumbling, then speaking aloud, a ten-minute torrent of words describing the epiphany he had just experienced.

"I know that the simplicity we sought in the sixties is not enough. I want to work at something where

I will be well paid. I don't lust after a lavish life-style, but money can be used to accomplish things of great value and produce great satisfaction. And money provides the freedom to travel, to live as we please, to experience life fully. Accomplishment and satisfaction are what I want. The clichés have fallen away and I can clearly see that purity and simplicity does not contradict working hard, being paid, and finding satisfaction in what is produced. It's all within my own power. It's my responsibility to do these things."

Mark was sitting on the concrete porch of the cottage in the early morning sunlight when Dale stepped out of his room yawning.

"What are you mumbling about?"

Mark shook his head, "I went up the path a little way to watch the sunrise on the mountains."

"How was it?"

"Insightful."

The painted bus jolted down the mountain to Katmandu square. Mark and Dale got off in the shadow of Bhimsol tower.

"I saw you outside in the moonlight last night," Dale said. "You sneak into that German girl's room?" He snickered.

"No, I was just looking at the mountains in the moonlight, " Mark grinned and added, "Past the broken fence I could see Ronald Colman crossing the snow, looking for Shangri-La."

"What the hell are you talking about?" Dale snorted.

They walked down the now-familiar alleys to the American Cafe, took a seat by the window, and ordered pepperoni pizza, and a Coke for Dale.

"Hey, man. Look at this." Dale said. "I even got ice in my Coke."

"You'll have dysentery tomorrow, from the dirty water they made that ice out of," Mark said and Dale hesitated. Mark shook his head. "Not to worry. Keep enough alcohol in your stomach and it will kill the bugs. We'll have a beer in a minute and you'll be fine."

Outside the restaurant, a Nepalese hawked up a wad of phlegm and spit it.

Mark and Dale burst out laughing. "The Nepalese national anthem," Dale said.

They walked down Sukra path toward Durbar square.

"I've been meaning to ask you, what the hell does 'roll out' mean?" Mark said.

"In racquetball, when you serve or return a shot so low and hard it just hits the backboard and rolls out along the floor its called a roll-out. Impossible to return. Learn that shot and you've mastered the game."

"No wonder you beat me every time we've played."

"You come to the club in North Platte after we get back to the States. I'll teach you." Dale fished in his pocket. "In the meantime, let's smoke a little of this hash."

In a narrow alley they leaned against the hand-made mud-brick wall and smoked a small bowl of hash. The occasional passing Nepalese ignored them. The hash glowing in his head, Mark led the way back out into the sunlight. A sunny courtyard appeared, a worn brick temple at its center, with steps on all four sides that led up to a sort of balcony at the top.

They explored the empty room at the top, but aside from dust and a few long-dead flowers, there was nothing there, so they sprawled on the sun-warmed steps and looked out across the city haze. The sounds of the city seemed very distant.

"Did you ever think you'd be here?" Mark asked.

"I never thought about going overseas at all," Dale said. "Let's go look at those tee shirts again. We're leaving tonight and I want to get one or two for souvenirs."

"You go ahead. I'm going to sit here for a while. I'll meet you at the hotel." Mark said. Dale got to his feet and went down the temple steps and disappeared in the crowd.

The still air and pale sun warmed Mark, bringing up memories of the Missouri University campus one spring in the late sixties. Flyers were circulating for 'Gentle Tuesday', lunch on the quadrangle, everybody invited, hippies and Greeks, SDS and Young Republicans, a change from confrontation and anti-war demonstrations. A Day of Love. He and Jennifer had walked through the crowd, someone was blowing bubbles, someone was playing a guitar, a Joan Baez song.

As Spring deepened toward Summer they spent every night together, listening to records and talking by candlelight, making love, smoking dope, under

238

posters that said Love and Hendrix and Joplin and Airplane and the Dead.

They read the political protest columns in the student newspaper, *The Maneater*, went to the sit-in at the Commons, and the one in front of the federal building on Cherry Street. They marched to protest the war and lay in the sun in Peace Park and talked about psychedelics and enlightenment, but they also studied.

Two American girls walked by on the square below. One had long black hair like Jennifer's "What we had back then was perfect," Mark whispered to himself. "I was twenty one and you were nineteen, we were in love, we were exploring art and history and literature and engineering and math and psychedelics, and music and philosophy. The times were changing, the world was changing, and we were right in the heart of it."

But now, Mark thought, that time has passed. I want to work hard, accomplish things, travel, do something satisfying with my life.

The bricks in the wall were pale red and yellow. Out over the roofs of the city, the air was hazy in the afternoon light. "I want absolute freedom, now, at this age, when the whole world lies before me." The sun shone down brightly and the people below seemed very far away. After a while Mark got up and made his way back to the hotel.

That evening Mark and Dale ate curry at the American Cafe. As they ate, Dale chattered away about the center-pivot irrigation systems all the farms used in Nebraska.

"You're pumping the aquifer down faster than Mother Nature can refill it," Mark said. "Pretty soon farming will be impossible. It will go back to prairie, then desert. Can't even live there like the Indians did since there won't be any surface water at all."

"Got to make money," Dale shrugged, "Can't farm the prairie without water."

"Well, maybe you shouldn't be farming the prairie."

"My buddy's started selling center-pivot irrigation systems in Algeria," Dale continued. "When my contract in Tabuk's over, I'm thinking about working for him for a while. I wouldn't mind spending a few months in Algeria, with a vacation in Europe on my way over and back."

Dale doesn't belong in Algeria any more than he belongs in Saudi Arabia or Thailand, Mark thought. He needs to be back in his hometown in Nebraska driving his Chevy Blazer with the big tires and the turbocharged 350 engine, playing racquetball at the club, dating girls at the country club, and going to the pool on a hot Saturday afternoon. That's his world. He feels uncomfortable in Thailand with all the dope and the drinks, the hookers, the food. It's all too much, too easy, it's intimidating!

After a while they walked back to the hotel, checked out, and took a taxi to the airport for the evening flight to Bangkok.

Outside the airplane window Mark could see the ghostly glow of the Himalayas in the distance.

At Bangkok airport they took the limousine bus to Pattaya Beach. When they arrived at the Orchid Inn at two in the morning, there was no Tom Farris registered there. "To hell with it," Mark said. "It's the middle of the night. Let's just get a room here for tonight. We'll find Farris tomorrow." They got two separate rooms.

The next morning Mark woke completely disoriented. He slid out of bed and drew the shade back a little. Nothing moved in the alley outside except a mangy black-and-yellow cat slinking purposefully down the cracked asphalt and into a house courtyard where the wrought steel gates had been bent. Mark went to the bathroom, then back to bed. The heat woke him. His watch said it was ten. He got up and turned on one of the floor fans which moved the humid air around. They found Tom on a chaise lounge under a blue and white umbrella at the beach in front of the Holiday Inn.

"Thought you were going to meet us at the Orchid Inn?" Mark said.

Farris shrugged, expressionless behind his sunglasses. Mark and Dale exchanged glances, then lay down on the closest chaise lounges, and spent the rest of the afternoon indulging in beer and shrimp cocktail. That evening the three of them were back at the Marine bar.

"Isn't that your girlfriend?" Tom said.

"I'll be damned. It is," smiled Mark.

Toy came over. "I look for you long time."

Mark put his arms around her. Her warmth felt good. This was the life, an absolutely free life. No

241

cares. He could choose any girl in the bar and have sex with her. Choose any drink and drink it. After a while they rode the baht bus, the local bus named after Thai currency, the baht, to the CouCou Lounge. Inside, a Brit disk jockey was playing Donna Summers. The dance floor was crowded. After an hour Mark took Toy with him back to the room at the Orchid Inn. He didn't notice when she left.

The next day, Mark and Tom rented motorcycles, big four cylinder Kawasaki 1200's, and roared northward out of town. At the first traffic circle, Mark followed Tom down the road toward Chon Buri. They wore no helmets. At 75 miles per hour, dodging through heavy traffic, the sun and wind felt good. They'd already had breakfast, sex, two beers and smoked half a joint.

In the afternoon, they played with jet skis in the blood-warm water in front of the Siam Bayshore Hotel. After that, they retired to their room to drink a beer, smoke a joint and drop into sleep for a while... It was late afternoon when they roused themselves and went down to sit by the pool.

"Let's go down to the Siam Bayshore again," Dale said.

The evening was cooling into night, air the temperature of silk, the sky the colors of indigo over gold. Waves rippled the pool's surface. Mark could see their white-jacketed waiter picking his way through the chaise lounges, three Heinekens and three glasses on his tray.

"Chase down some round-eyed women?" Tom asked.

Dale smiled his tongue-forward smile. "Sure. That's where the Qantas crews stay."

"Sounds interesting. Or to go from the sublime to the ridiculous, we could cruise the Marine Bar again, or that disco. What was the name of it?" Mark asked.

"CouCou Lounge. I think your girl was embarrassed you were still dressed in your swimming suit and that surf shop tee shirt last night at the disco," Dale said.

"You didn't look much better," Mark said. "Covered with dirt from climbing on top of that baht bus we were riding back to the hotel."

Tom signed for the beers and laid a 20-baht note on the waiter's tray.

"Thank you, sir." The waiter bowed, taking the money and shoving it in his pocket.

Mark took the cold Heineken in his hand, sipped it, then slowly set it on the white plastic table beside his deck chair.

He closed his eyes. The slapping of the plastic float in the pool was subtly different from the ones in the compound pool back at Tabuk. And no flies here. He visualized the white heat over the compound. The smell and sound of the cafeteria, the Palestinean waiters, each wearing their *kaffea* at a slightly different angle. He pictured the people he knew going about their daily tasks – Jim, Dick, Mr. Pak, and all the others..

Ripples shook the lane stripes of the swimming pool in the fading light, the windows of the hotel were gold.

"I'm sated," Mark said quietly, "Satiated with food, sex, drinks, but I'm not quite satisfied.

"Drugs, sex, and rock and roll," Dale grinned skeptically.

Mark took another minute sip from the sweating green bottle.

"So are we going to chase round eyes tonight, or retire in civilized splendor to the Swiss pork chop place followed by dancing, drinking, and whore chasing at the Marine Bar?" asked Dale.

"That's a tough choice."

"But the first step is obvious," Dale said. "We go up to the room and blow a little smoke."

So they did.

But it felt cold and sterile in the cool, quiet eleventh-floor Holiday Inn room. One after another, they would get up and pace.

Finally the joint was gone.

"Well, where to next?" Mark said. "If you guys don't have any great ideas, I'm going to go find Toy back at the Marine Bar."

"Sure," Dale said through a lung full of smoke.

Mark was glad to get out of the little room filled with these two big guys. Tom's bare chest, his too-short cutoffs, his perfect teeth and perfect suntan. Dick's pacing and too-loud jock talk. Tom's over-confident heartiness. These guys are starting to irritate me, Mark thought. That night was the same as the previous night.

The next day at noon they were back in the chaise

lounges at the beach under blue-and white umbrellas. A waiter materialized.

"Three Kloster beers, please."

Dale walked down to the water and kicked some spray into the air. A tiny perfect surf rose and broke on the sand.

"This is paradise!" Dale smiled, perfect teeth in his broad suntanned jock's face. "I'm happy."

That afternoon they went jet skiing and parasailing and drank beer continuously. Toward evening, Tom volunteered. "I'm not happy."

"We know why not, so don't tell us. Let's go get some dinner." Mark stood up. The evening breeze was such a neutral temperature he couldn't feel it. Mark was thinking of Tabuk, the job site. He pulled himself back to the present. The other two guys were still lying there, sipping their beers.

"Hey, look, guys. I'll catch up with you later, okay?" Mark said.

"Where are you going?"

"I don't know, man, I'm just going to stroll around a little. Where are you guys going to be later?"

"CouCou lounge, most likely," Dale said.

"See you," Tom's smile showed all his big square Teddy Roosevelt teeth under sad eyes.

Mark walked briskly for half an hour, going nowhere in particular, just clearing his head. He went into a store at random. The bell on the door jangled as he closed it.

Thai music played softly in a back room. He smelled scents of sandalwood and Thai cigarettes.

Mark looked at the rolled Chinese carpets, the

carved teak wood from Malaysia and Pakistan, Toshiba cassette players. Two moths flickered around the bare light bulb in the ceiling.

He stood still in the restful atmosphere for a long time. Then the door opened behind him and two French women walked in.

He dismissed their unfriendly glance as they brushed by him and began pointing out some carvings to each other.

The tranquility broken, he went back out into the humidity and the noise and went up Bayside Drive, past the lobster aquarium in the front of the Surfside Restaurant to the first big bare bar.

There were a couple of Brits at the bar talking into each other's faces while their bar girls sat ignored behind them. The rest of the people in the place were Thai women, maybe twenty of them. After a while he went back to the hotel alone and went to sleep.

The next day they took a taxi to the airport and got in line at the Alia airlines counter. None of them were saying much and everybody else in line seemed equally subdued. Dale elbowed Mark in the ribs and tilted his head toward a blonde in line ahead of them.

Mark shrugged. As he followed Dale through the metal detector, he noticed the indicator lights were not glowing, the machine was not even switched on.

On the plane the three of them sat in a row, Mark by the window. Flying west, night came slowly. After dinner trays had been cleared away, Tom went to sleep. In the darkness over India Mark told Dale "I came to Tabuk looking for travel, for the responsibility of running a big project, for being completely free. Even of relationships."

"Looks to me like you jumped from one relationship to another. Most people want a relationship," Dale tilted his head toward Tom.

"A lot of the ones that are in long-term relationships sure don't seem very happy." Mark moved his empty Pepsi can around on his tray table. "I know I want a long-term relationship, eventually. It's just that I want to be free for a while first. It's like going to work at some job and spending your whole life working at that same job – that's what makes people bitter. Same with relationships."

"My father and brother are still back in North Platte farming that same six hundred acres," Dale said softly. "I'm glad I came to Tabuk. This will be something I'll tell my kids about someday, but I'm ready to go back to Nebraska." The dark plastic window reflected their faces, both serious.

Dale turned off his reading light, "I'm going to sleep."

Mark managed to slip into a half-awake stupor, the airline equivalent of sleep. He saw his Blazer with the wide tires parked in front of his little house back at Tabuk. In the cool bedroom, Cathy waited. I like all that, he thought. But it's not meant to last forever. In his mind, he fanned through the couples he knew like a deck of cards, Brian and Linda Zeller, Jim and Ti Redding, Tom and Trish Farris, Floyd and Danielle Calvin, and Allen and Carla Hayes back in the States.

The moon was a pale crescent in the blackness. I

want to get married, but not yet. I don't want to end up like Dick and Kurt, all alone when I'm fifty, sixty years old. But Jennifer and I met too soon. He stared at the blackness for a long time. "I'm sorry, Jennifer."

Eventually the plane began its slow descent through an orange dawn toward Amman airport.

Chapter 13

Teams of four soldiers with AK-47's were patrolling the airport when they landed in Amman. Four American made M60A3 tanks were parked in front of the airport. "What's this all about?" Dale muttered.

"Don't know, but let's don't stick around to find out," Mark said. They hurried to his dust-covered Blazer, drove past manned checkpoints and out of the city onto the desert highway. The border crossing into Saudi Arabia took more than two hours, lines of vehicles being inspected, Saudi troops, three separate passport checks.

Mark went to work as usual the next day and it was business as usual with the Koreans. They never discussed politics – under orders from their embassy. It was Wednesday, so after work Mark decided to go the weekly poker game. Redding told him it was at Brian's house. Brian's wife and kids had been gone for months, but Brian had whined so much that Vance had let him stay in married quarters.

"Surprised you're here, Mark," Brian said. He took the blue-and-white chips out of the cheap brown plastic chip holder with an imitation roulette wheel molded in the top. "Thought your domestic duties were keeping you occupied full time."

249

"I get parole once in a while." Mark was getting real tired of the jokes.

Kurt wandered out of the kitchen, a generous glass of scotch in his hand.

"Help yourself, Kurt," said Brian sarcastically.

"I will," Kurt said with a big grin. Brian cashed riyals for chips for everyone and the game got underway.

"What the hell's going on at the border anyway?" Mark asked.

"You haven't heard?" Brian the busybody said. "Soviets invaded Afghanistan. Four mechanized divisions. Took over the whole country in four days. Vance called an emergency meeting yesterday, briefed us all."

Dick poured himself another drink, "Well, Soviet troops, tanks, air superiority, artillery, all that, no wonder they made quick work of a bunch of Afghan tribesmen."

"But whether they can hold on to what they've got," Redding said. "That's a different story. The Vietnamese beat us in Vietnam, we had all that technological superiority too."

"The surprising thing," Kurt said, "is that the Russian troops were conscripts from Tajik and Uzbek." His Bavarian accent had thickened with the scotch. "I'm surprised the Russians took that chance, but the invasion itself is no surprise. The Russians don't want the Muslim fundamentalist revolution that's started in Iran to spread to the Muslim populations in south central USSR." Kurt set three chips neatly in the pot, laid out two pair, and raked in the pot.

"The only people surprised by this are you Americans. You people just don't seem to have any grasp of world politics."

"Let's don't talk politics," Dick said.

"U.S. foreign policy is a joke," Mark said. "Especially this human rights campaign Carter has been pushing. It only alienates foreign governments, pushes them over to the Soviet side. Totally ineffective. He can't even get the Russians to the negotiating table for SALT II."

Cards were dealt.

"Well, our State Department is right to help this guy who took over Iraq, Saddam Hussein," Ray Barton said adding chips to the pot. He laid down a full house and swept up the pot. "State Department is giving Iraq all the military supplies it wants because Hussein says he's going to invade Iran, start another holy war."

"Hey, would you guys knock off the politics," Dick slurred.

"And we're supporting that?" Mark continued. "You'd think President Carter's advisors would be smart enough to see these local dictators are totally unreliable. For us one day, against us the next."

Kurt laughed long and loud. "Most American government people have never been out of the U.S.A. They have no idea what's going on anywhere except back home in America."

"Hey guys," Vance said. "Let's play poker and forget the political discussion."

Mark lay on his back in the gray of predawn staring at the strip-boards that covered the joints between the sheets of fiberboard forming the ceiling. He'd turned the window AC off an hour ago. In the States there would have been the sound of birds, of breeze in elm and oak leaves, crickets, the occasional passing of a car, or someone's lawn mower.

The stillness of a desert morning was absolute.

Friday morning, Al Jumwah, the day of rest. The holy day.

Mark turned his head to one side seeking a cooler spot on the pillow, being careful not to nudge Cathy. Why doesn't she do something with her hair? The way she cuts it, and its dull blonde color always looks like it needs washing. If it were cut shorter and whiter, that would be sexy. He slipped out of bed quickly before she woke.

Mark pushed the button on the face of the flash water heater beside the shower and let the low pressure water trickle over him. He remembered the dim blue vastness of the ocean in his dream. Toweling himself off with one of his two threadbare towels, he made up his mind to go out to the job site for an hour or so.

Cathy was sitting on the couch in her robe. She smiled as he crossed to the bedroom to get dressed.

"Good morning," she said.

He grunted and pulled on his jeans, one of his old knit golf shirts and slipped on his desert boots.

"I have to go out to the job for a little while," he said lamely.

"Friday? You don't even have one day off?"

"It won't take long. A couple of hours."

He finished lacing his shoes. He wished the house had a back door he could slip out, without seeing her or hearing her. He stood silently and opened the drape a little. The sky over the backyard wall was white. The glass in the window already felt warm to the touch.

"Want some breakfast before you go?" She said. He heard her open one of the kitchen cabinets.

"No, thanks."

He forced himself to go out to the kitchen. She had the refrigerator door open and was eyeing the four remaining strips of bacon. He put his arms around her. "You go ahead. Fix yourself a leisurely breakfast. I'll be back before you know it." He kissed her and left quickly.

Outside, the day was hot and still and bright. Flies buzzed and he swiped them away as he got the Blazer's door open and hoisted himself inside. He started it up, adjusted the air-conditioner vents and pulled slowly down the line of tamarisks to the compound's main street.

"Good morning, sir," Mr. Lee greeted him deferentially at the Dorm project.

"*Ani Hashimnika*" Mark answered. He left the drawings spread out on his desk.

"Can you make inspection?" Lee inquired politely.

"Where?"

Lee handed him a slip of paper, a form the company had devised, requesting a leak test final inspection at some drain lines on the third floor.

"Ready now?" Mark asked.

"Any time."

"Where's Mr. Kang?"

"Not on job site. Korea."

Mark studied the Quality Control organization chart behind his desk. Chief of Quality Control was Mr. Kang. Chief Mechanical QC was Mr. Kim, but he was worthless. Mr. Pak was in charge of shop drawings but was a good field inspector and a lot less argumentative than Kim.

"How about Mr. Pak? *Pak Ja Su*."

Lee stared at him.

"Bring Mr. Pak. *Pak Ja Su*. Shop drawing."

"Shop drawing?"

Mark got up, got his hard hat and tape measure, held the door open for Mr. Lee. They walked over to the Dae Joon quality control office.

Mr. Pak was talking with a couple of the young Korean engineers.

"Good morning, sir." They stood and bowed as Mark approached.

"Morning, Mr. Pak. Ready for inspection?"

Lee and Pak exchanged a few words. Mark had them open the P-sheets in the contract drawings and show him the location and extent of the piping under test, then the shop drawings showing the exact layout.

The shop drawings the Koreans drew were beautiful, sometimes better than the contract drawings. Back in the States, most contractors and their mechanical subs refused to make shop drawings. Even if they had tried, drawings of this quality would have been far beyond the capabilities of the crusty old plumbers and unskilled laborers that comprised American crews. Here, all the quality control staff were graduate engineers.

Bad parenting and bad teaching will ensure the end of U.S. competitiveness against more motivated and better educated Asian workforces. We'll soon be living in an Asian world. Guess I'd better enjoy these days while I can.

Mark flipped back and forth from the architectural sheet that showed the bathroom dimensions and fixture layout to the plumbing sheets that showed the piping size and locations.

"I need to see lavatory and toilet submittal," Mark told Pak.

Pak snapped an order and one of the junior engineers rushed off. When he had them, he flipped back IIAA form number 4288 on the front and copied the rough-in dimensions from the American Standard drawing. Distance from toilet back to drain centerline was 30 cm. It needed 60 cm on either side of drain centerline for installation and clearance.

Mark noted the route of the 4" drainage piping and closed the drawings.

At the third floor, Mark stepped up on the box beside the ten-foot standpipe. His tape measured five cm down from the top of the pipe to the top of the water. He noted that and the time on a piece of paper and stuck it in his jeans pocket.

He and Pak and their entourage made their way down the ten rooms on this wing checking pipe stub-out locations. Mark was subtracting 2 cm from the 60 cm lateral clearance requirement to compensate for wall sheetrock, setting bed, and ceramic tile which was not yet installed.

They finished the dimensional check in twenty minutes.

"Slope?"

"Sure."

They all clomped down the dark stairwell. "Where are the temporary lights?"

Pak exchanged words with one of his engineers who went off shouting at the nearest laborers. Mark kept moving. On the second floor, he measured the distance from the top of pipe to the bottom of the concrete from which it was suspended, then the same at each of the row of rooms. The measurements showed a steady 1/4" per foot drop in accordance with the specs.

"They're using service weight hub-type cast iron. It's already in the ground." Redding lit another Marlboro. "Got a set of drawings?"

Mark brought over a set of half-size.

"You can't see anything on those. Let's look at full size."

Mark thumbed through the drawing rack under the clattering window AC.

"Which building?"

"3121. The pump house."

Mark pulled a heavy set of drawings off the rack and laid them on his desk. He flipped back to the P-sheets. Redding showed him the line.

Mark studied the line, then the drawing notes.

"Nothing here. Let's take a look at the specs."

Redding had volume three open. "Section 15401, right?"

"Yeah. About paragraph 6 or 8. Somewhere in there."

"...below grade Drain-Waste-Vent piping shall be extra heavy cast iron in buildings of two or more stories," Redding quoted. "That's a one-story building."

"Then service weight's okay?"

"What's the difference between the two? Just wall thickness of the pipe?"

"Yeah." Mark flipped drawing sheets back to the exterior utility sheets. "You know what may be a problem, though, is where the line coming out of the pump house connects to the line coming out of the main building. The main building line is going to be extra heavy."

"It's already in place. That was before you got here. Last summer. Preece looked at it."

Mark nodded. "I guess I can mark those lines off then." He gestured toward his desk. "I'm keeping a record of all the lines tested. To be sure we don't overlook any."

Redding stubbed out his cigarette. "Exterior lines are lined Transite. So the service and extra heavy will already be transitioned to Transite."

"I doubt it. Even though we're out past the 5-foot line, they'll keep going with cast iron. The pump house line goes under the main building and connects to the main line there."

"That's a shitty design," Redding said. "They

should have had separate drains. If you get one line clogged, you've got both building drains out of action."

"Can't. That's all paved driveways between the two buildings. Not a good idea to run parallel Transite lines under all that pavement. If you ever had to get to them you'd have to tear up half the driveway."

Redding took his hard hat off and shook another cigarette out of his Marlboro pack. "So what should we do? Is there a fitting to tie service weight to extra heavy?"

"I'm sure there is. But they're not going to have it. They'd have to order it from the States. Even airfreight it would be three or four days."

"I can't wait that long. We've got to get these trenches closed up so we can get cranes in there to do the roof frames. That stuff's on the critical path."

Mark studied the drawings. "How about this? We let them run service weight out of the pump house, under the main building to the main drain. That stuff over there, under the main building mechanical room, won't be done for weeks. That'll give them time to order the right fitting."

"You're letting them run service weight under a multi-story building. Violates the specs."

"Yeah, but there are no risers to it, it's just handling single-story flow. Functionally, it's fine. So what I'll do is tell them that we will let them run service weight under here, but that they have to add a floor clean out right about here."

Redding studied the drawing. "That's carpeted

floor."

"Okay. So we'll put it just outside the building. Better access there anyway. And there's a little strip of turf around the buildings. Pavement doesn't run right up to the walls, does it?"

Redding studied the paving sheet details. "There's a thirty-centimeter landscaped strip between the building wall and the pavement curb. Can you get a clean out in there?"

"Sure."

"Talk to QC about this stuff and be sure they write it down in today's QC report. I don't want them to conveniently forget that extra clean out in a week or two."

Mark closed the full size drawings and put them back on the rack. "I'd rather just have them mark up the as-builts."

"Do it today."

Mark finished up and drove to the Kendall company mess hall at lunchtime. All the guys were sitting staring at their coffee cups.

"What's the matter with you guys?"

"IIAA is closing the office here. It's on the bulletin board outside Vance's office. Manpower will be reduced, and this site will be managed by staff at the Jidda office."

"When the end comes, it comes fast," Dick said. "Things are going along, then things change, and next thing you know you're the last one standing."

"Well, turn out the lights when you leave." Danny scuffed his boots under the table.

"It's funny how things can change," Ken Cooley said. "You go along for months, years, just focused on the job, on your friends and coworkers. The big world outside doesn't exist."

"And then all of a sudden it does exist," Barton said.

"Just like Vietnam," Jim Redding said.

"And Pakistan," Dick added.

"And Iran way back when," Mike Robb, the old-timer said.

"If you transfer to Jidda, you taking your girlfriend with you?" Danny asked Mark.

"Well, I..." Mark stuttered.

Dick smiled his creamy smile, "We sure get nosy don't we?"

Petri burst into the mess hall, "Vance is calling an emergency meeting right now. Everybody over to the office."

At the office, all the chairs were full and guys were standing around the walls of the room. It was already too hot.

Vance came in with some papers in his hand. He smiled. "You've already heard all the rumors, so let me give you the facts." He thumbed through the papers. "The news services are reporting that a large mob, mostly students, have stormed and captured the U.S. Embassy in Tehran." A rumble of conversation started. Vance held up his hands, "Listen up! It is being reported that a large number, maybe one hundred American employees and dependents in the embassy have been taken hostage. The kidnappers are demanding the extradition of the Shah back to

Iran to stand trial..."

"...And be executed," several people said at once.

"President Carter is to release a statement at noon Washington time. That would be about 6 p.m. here."

"Our dependents going to be evacuated?" Barton asked.

Vance looked around the crowded room. "There is no danger to us here in Saudi Arabia, not at this time. So it's business as usual. Just be on your toes. Nobody is to leave the compound until further notice except to go to the job sites and return directly after work. All kids and wives to stay inside the compound."

Brian stood up, red-faced, "Yeah well, this is bullshit. I'm ready to leave now. Air Force can bring a C141 down from Germany today."

"Don't get panicky," Vance said. Across the room Redding rolled his eyes at Mark.

Vance continued, "The State Department has notified all American agencies and companies doing business here that large-scale evacuations are not approved since it will send signals to the Saudi government that the U.S. is not supportive of the Saudi government."

"Damn good thing Saddam Hussein took charge in Iraq," Ray Barton said. "He'll stand up to any post-Shah government in Iran."

"And none of that talk!" Vance snapped. Barton's head jerked back as though Vance had slapped him. Vance swung his gaze around the room. "It is inappropriate for American nationals to be making comments on inter-arab relations. Especially now."

"President Carter and Secretary of State Cyrus Vance have been praising Saddam Hussein in the newspapers," Hager said. "That's the official U.S. position, it ain't just us saying it."

"Danny, that's official State Department business. Our business is construction management, so we keep our mouths shut."

"Has Khomeini returned to Tehran yet?" somebody asked.

Vance thumbed his sheaf of papers, "The news services in Paris say the Ayatollah is supportive of what he is calling the 'people's revolution', but he has made no comments on the hostage situation and there is no report of him planning to return to Tehran."

There was a lot of talk, mostly about getting women and children out. Vance said he'd request a special flight from the IIAA office in Jidda.

The next morning, Mark took Cathy out to the desert for a scenic drive.

The Blazer jounced slowly over red rock at the edge of the silent valley. Cathy passed Mark another cup of lemonade heavily laced with vodka. They started down the rocky hillside toward the tire tracks in the sand at the bottom of the valley. "Let's don't talk about politics, or us, OK?" Mark said.

Cathy recapped the thermos and put it on the floor. "Alright. I thought we weren't supposed to leave the compound."

"It's safe out here in the desert." Mark said.

"What do you think we'll find beyond this valley?" Cathy asked.

"Another valley. And another."

"And more rock carvings?"

"Perhaps."

"Mary told me they were ancient. Dating back to before Mohammed."

Mark steered the Blazer around the last rock slide and accelerated as he reached the loose sand. "You might want to hang on to something for the next minute or two. I've got to keep my speed up or we'll get stuck in this sand."

After a couple of hundred meters, they drove up onto gravelly hardpan.

Where the trace of roadway wound into the dusty red, gray, green and purple boulders, Mark spotted some marks on a rock.

"There. See that?" He slowed the Blazer to a stop and pointed. "Evidence of the ancient Nabateans. The spice caravans used to come up the western coast of Arabia, right through these valleys, on their way to the Phoenician cities of Tyre and Joppa. Caravans of myrrh and coffee from Mocha, frankincense from the Sudan, gold from King Solomon's mines in Ophir."

Cathy studied the markings on the rocks, her head bent low to look up through the windshield. The air conditioning blew her blonde hair all around her head.

He touched her hair, felt his arousal, he unfastened his seat belt and slid over to her. He kissed her, felt her breasts, unfastened her bra and ran his tongue over a nipple.

"Oh, Mark..."

They made love in the car with the desert all around, bare and hot and empty. Back at his house, he tossed his sweaty tee shirt on the floor and dived into the rumpled bed and pulled her to him again. The blue glow from the curtained window and the hum of the window air conditioner blanketed him in a pool of darkness away from the midday glare.

In his dream, he and Jennifer were driving down two lane blacktop in his old Chevy. It was late summer, a thunderstorm was building above the rolling hills of oak and elm and sycamore. Cathy's sniffing woke him. She lay with her face turned away from him.

"Are you getting a cold?"

"Sorry."

Sleep began to rise over him again. In the darkness her words were so soft he could barely hear them.

"Please stay with me, Mark."

He said nothing. I wish it was cold, he thought. I wish I was far away in some cold and snowy mountain cabin deep in a pine forest.

"I got my ticket home today," Cathy whispered. The delicious darkness pulled at him, but the colors had gone out of the dream.

"When?" he asked.

"Day after tomorrow."

The night was dark and motionless. Mark looked at the outline of the doorway and the gray light from the living room.

"Will you drive me to the airport in Amman?" she

whispered.

"Sure."

Mark felt his heart beating slowly, like an engine under load.

Mark got out of the Blazer and walked the job for twenty minutes. He walked through the nearly-complete first floor, through the second, third, and fourth. The Koreans saluted as he passed. He nodded in return and kept moving.

Mr. Lee spotted him on the fourth floor and bustled over but Mark brushed him off.

After some time, he drove in to the IIAA office.

"Hi, Ursula." He sorted through the official correspondence in the project office basket.

"There's something for you in there."

The little spy, he thought. I'll bet she reads everything that passes through this office.

"Good news, I hope," she pried further.

Mark paused to read a letter from Dae Joon Company asking for official clarification of the arrangement of air handlers in the mechanical penthouse. He thought he had worked that out with Mr. Kim the mechanical QC. I think maybe Mr. Pak, the Dormitory building manager, is putting pressure on him. I think he's out of favor. Maybe they are building a case for sending him back to Korea. The senior engineers were expected to work everything out themselves or in person with the IIAA guys. Even the shop drawing submittals were discussed in advance. To send a letter after discussing an issue was something only an American company would do, very legalistic, and very out of character for the Koreans.

"Can you log this in, Ursula?" Mark handed her the letter. She had some sort of elaborate logging system in a notebook she used to keep track of all the office correspondence.

Mark found an official letter addressed to him from the IIAA personnel office in Jidda.

"Thanks, Ursula, see you." He went out to the Blazer to read it, away from Ursula's snooping.

It was a job offer, at IIAA's big new Riyadh hospital expansion project. He needed to let them know one way or the other by the 19th.

He went back to the job site and worked until lunchtime. When he came back to the house after lunch, Cathy had gone to work. He got the job offer out of the envelope. Then, without really thinking much about it, he marked "accept," signed and dated it, drove back to office and got a routing envelope from Ursula. He wrote IIAA's personnel office symbol and slid the job offer inside.

"Oh, I need to make a copy of this," Mark said.

Ursula eyed him. "I'll do it."

He reluctantly handed the paper over. She made a great show of not reading the form as she made a copy on the creaking Canon copier behind her desk. He was sure her beady little eyes had scanned the form. By this evening, everyone in camp would know.

Chapter 14

The final Wednesday night poker game was at Mark's house.

"Where you going, little buddy?" Dick slurred, "This our last game, don't want to miss it."

"The smoke in here is so thick I can't see my cards," Mark said. "I'm going to open the door."

"That's okay. Just so we can see your cards." Ray said "Any more scotch?"

"On the shelf in the closet."

Mark dealt the cards, looked at his hand and folded.

Redding tilted his head "Guy I used to work with at Vinnell in Da Nang used to say you either kissed the past good-bye or you kissed your future good-bye."

Mark finished his beer and stepped outside. In the light he could see that the grass Cathy had tried so hard to grow was dying. The hazy sky revealed only a few dim stars. He leaned against the warm concrete block wall and looked out toward where the flat horizon was invisible in darkness.

"Hey, Mark, you going to play or not?"

"Yeah, not this hand, though. Give me a minute."

He missed seeing the stars. Orion high in the sky

on a cold night in January driving over to pick up Jennifer at her dorm for a Friday night date.

"Hey, Mark, you playing or not?"

"Let him stand out there. That's okay," Dick's voice rose. "We're contributing five of your riyals to the pot. Taxes."

Mark remembered an evening with Jennifer at his apartment in Columbia. She'd flipped open a book she was reading for her literature class and read him a passage. "Hemingway stayed in love with his first wife his whole life," Jennifer said. "He divorced her, but stayed in love with her." She looked at Mark, "Maybe he just didn't realize all that until a long time later. Here's what he wrote her in a letter, "...maybe we will have a fine time together in heaven, and maybe we have already had the hereafter and it was up in the Dolomites, and the Black Forest, and the forest of the Irati. Well, if that is so, it's okay with me..." Sometimes we don't realize how good the times are until they are long past."

Mark came back to the present, glanced at the hazy night sky and went back in to the poker game.

Dick was proposing a toast. He raised his glass. "Well, here's to us all, may we meet again." They all raised their glasses and drank.

"That's the way it is in construction," Dick said. "Projects end and people move on to the next one."

And that makes it easier, Mark realized, where everything has a clear beginning and a clear ending, projects get finished, people come into your life, and they leave. It's less ambiguous. Back in the States, we let things go on and on, even though we know they are not good, just because it's too much trouble to change them.

The next morning Mark got up early and drove over to the HSC dorm to pick up Cathy. She had moved all her things out of his house the week before and had been living in her dorm room since. Mark went in the gate and up the stairs and found her sitting on the bed in her room, bags packed.

They drove down the smooth new highway while the mountains and desert glided by like a dream. Cathy dozed against Mark's side. The familiar desert valleys drifted by. Along the highway were sun-bleached mountains of pale red, green, and purple boulders like the backs of great sleeping lizards.

Finally they crossed the last mountain pass and started down toward the Gulf of Aqaba lying blue and sparkling before them. The desert had disappeared behind him silently, and his year in Tabuk had become the past, a landscape on which to paint the rest of his life.

Mark was thinking of the green trees of Missouri. He'd be sitting in the lawn chairs under the walnut tree back of Allen and Carla's house in three more days.

Security at the Jordanian border was very tight. It took nearly an hour to get past the searches and the paperwork. They finally got to the Holiday Inn at Aqaba and checked in as usual. They walked down to the nearest shops and wandered around for a while trying to enjoy the day, but that was difficult, knowing what was coming. Later they sat on deck chairs on the little beach in front of the swimming pool drinking Amstel beers. The sinking sun sparkled on the cold blue water. Desert rose from the shore without any intermediate fringe of greenery. They talked aimlessly for a while. When the sun touched the tops of the hills behind Eilat on the Israeli side, they went back to their room, cleaned up and went out for a very subdued dinner. Later, they tried to make love, but their hearts weren't in it. Finally they just lay quietly, holding each other.

Mark saw a tear slide out of her eye and silently roll down her cheek. In a moment she was shaking, trying to suppress her sobs.

"Mark, Mark, please...I love you. I don't want to leave you.." She pushed closer to him, put her head against his shoulder.

She rocked and patted him as though he was an infant. "Please, please..."

Unable to respond without seeming insensitive, Mark remained quiet. Eventually they both managed to doze for a while, hiding in sleep, exhausted.

The next morning they drove up to the airport. Mark sat with her until the flight to New York was called. They waved to each other. After the plane had gone, Mark drove back to Tabuk keeping his mind empty. It was dusk when he parked the Blazer in front of his house and went inside and turned on the light. It felt very empty.

At the job site the next day, the Koreans had white paper laid out on the table weighted with little pyramids of four oranges. At each end of the table was a stack of glasses and little clusters of Sohat water and

Moussy near-beer bottles.

Mr. Kim saluted. "Small party."

Redding showed up and was handed a Moussy. He kept his hard hat on even inside, same as always. "Hear you're going to stay with IIAA," he told Mark.

Mark nodded. "Yeah, got my paperwork the other day. Hospital project in Riyadh."

"Good for you. You do good work."

Mark grinned and they shook hands. "Thanks." He glanced around the room. The Koreans were talking among themselves deferentially. Redding gave a small speech, thanking the Koreans and looking forward to working with them again on another project. Mark said a few words. The Koreans applauded and bowed in unison.

"I appreciate all the help you've given me, Jim. I've learned a lot," Mark said to Redding on his way out. Jim grinned around his Marlboro. "Well, we can all use some help. Except Preece, who already knows it all."

Mark had told Dale he'd say goodbye to him at the HSC recreation center. It occurred to Mark he had never been invited into Dale's room, although Dale had been to his house dozens of times. Dale was waiting in the entryway. They shook hands and Dale walked him out to his Blazer.

He stuck out his hand and gave Mark another over-strong handshake. "Well, be seeing you."

"Your place in North Platte next October, right?"

"Right. I'll get you a free game at the racquetball

club."

"Yeah," Mark said. "Roll out."

"Right. Keep in touch."

"You too," Mark said, knowing they never would.

Mark parked his Blazer at the edge of the road that led to the fuel facility. A Dae Joon grader was doing the final trim on the gravel road. All the scaffolding, the job shack, the fence, and the signs had been removed; the facility was complete. At the end of the neat roadway, the white painted tanks and piping gleamed in the noonday glare. Mark sat there letting the Blazer idle, the air conditioning running.

The completed facility looked good. Pride flowed through him but it was not a selfish pride, but a pride driven by awareness of his own ability to take responsibility for people and money and to judiciously exercise authority. It had to do with being part of something larger than himself, of becoming an accepted member of the brotherhood of those who accomplished things. A brotherhood based entirely on proven results, not talk. After a while he turned the Blazer around and drove back to the compound.

On the way, he stopped at the nearly completed Dormitory, parked in the newly paved parking lot and walked through the lobby. The air conditioning was on, the tile was polished, the scent of paint and new sealant filled the air. Everything was clean and beautiful. He went into the theater-style conference room, walked halfway down the sloping aisle and took a seat. A feeling of complete relaxation came over him. He rubbed his hand on the desktop in the armrest. He looked around the theater, thinking of all the hours and days he had studied the design drawings, inspected the work, and resolved the problems. A year of his life had been spent here. He felt a wordless power inside, a subtle elation, the sense of competence and completion. And I know where all the flaws are, the mistakes we corrected, the changes we made. He walked out the door and closed it carefully behind him. Now it belonged to someone else.

He drove through the IIAA compound, went to his house, got his swimming suit and a towel and drove out the gate and along the desert road through the empty valley beyond. He eased the Blazer over the flinty road, sped along through the soft sand, desert driving all second nature to him now.

After driving through the valley where he and Cathy had made love on their last pleasant desert outing, Mark stopped the car and powered the window down. It was absolutely silent under the white sky. The confidence and elation he had felt walking through the completed Dormitory building left him. I should have treated Cathy better, he thought. I tried to be honest with her and with myself, but I know I hurt her. But I don't know what I could have done differently without causing even more pain. We are just at different places in our lives, going in different directions.

He kept going, out of the valley a few kilometers to where the desert road met the paved road to Aqaba. He drove to the crossroads and turned off on the sand road that led to the coast. Where the desert came right to the cold blue waters of the Red Sea, he parked the car, slipped into his swimming suit, got his rock shoes and mask and snorkel out, put them on, and splashed out into the water.

At the edge of the reef wall, he dived out into the cold, deep water. He drifted along the reef face in silence, watching the kaleidoscope of fish and coral. Below him the clear water hazed away into blue nothingness. Removing his snorkel from his his mouth, Mark floated on his back, looking at the empty white sky. Still on his back, he began an easy backstroke, straight out into the sea. He looked back and saw that his car was just visible in the white line of the desert. He was a half kilometer off shore, perhaps. He rolled onto his belly and floated, looking down into empty blue. On all sides was only empty blue. He took a breath and dove down into the cold water, then let himself drift slowly up to the surface. He did it again. The fear was gone now.

After a time, he swam back to shore and clambered out past the scorpion fish, and dried off. The sun quickly baked the chill out of him, and in a minute the flies were buzzing, so he changed back into his clothes and started back to Tabuk

I'm not sure where I'm going, he thought, but I'm not going back to what I was. The road scrolled by, empty of traffic under an empty white sky.

This colorless emotion is like good health, when you have it, you feel nothing, but when you don't have it, you know something's missing.

At the crossroads, on impulse he turned toward Tabuk town instead of going directly to the compound. "I need to get some kind of souvenir for Allen and

Carla." The ubiquitous Toyota pickup trucks were parked everywhere, but there was nobody in sight. He parked in an alley back of a row of shops, and walked around in front. Loudspeakers were blaring in the town square two blocks away. The corrugated iron shutters were down on all the stores. Prayer call already? The loudspeakers continued their sonorous pronouncements.

He noticed all the shops up and down the street were closed. On impulse he walked to the square. It was packed with arabs, a mix of town folk and bedouins. They were all standing silent. Mark craned over heads and saw six men kneeling in front of a cleric reading into the microphone. Mark's heart started pounding. There were ten guards in khaki Saudi National Guard uniform and a small squad of Royal Guards in white and red. A man at the side had an axe. There was a large wooden block in front of the six men.

The cleric stopped reading, looked at the crowd, and nodded to the man with the axe. A prisoner was pushed forward, his head put on the block. The axe arced down and a head tumbled to the sand. Mark turned and elbowed his way through the crowd and back to his Blazer. He drove carefully out of town and back to the IIAA compound, emptying his mind of the scene he had just witnessed.

At the gate were two U.S. Army military police checking IDs. Mark drove to his house and sat in his living room staring at the white wall, then drove to the mess hall.

275

At the mess hall, nobody was saying much.

"Brian is gone. Got a seat on an Air Force flight to Germany yesterday. IIAA flew Ursula, Beth, the other wives to Riyadh," Dick said. "Except Del."

"She's getting office stuff and household goods packed and shipped."

"When are you out of here, Mark?"

"Tomorrow."

At the motel, two Palestinian houseboys were furiously vacuuming the big persian carpets in the lounge under Del's critical eye. Her hands never strayed far from the pack of Marlboros on the desk in front of her.

"How much household goods will you have?" She slid her ashtray to one side and began filling out a form. Mark could barely hear her with the racket from the vacuum cleaners.

"Not much. Two suitcases. Plus a couple of boxes of papers I'd like sent to the office in Riyadh."

"You're entitled to five thousand pounds." She wrinkled her West Texas complexion into a squint. "You're still single, right?"

"Yes, Del, I'm single. I'm not taking much more than I brought here in the first place. I didn't load up with souvenirs like some others."

She gave him another look. "Travel light, do you?" She nodded to herself, equating the paucity of his belongings with the paucity of his life.

She opened her key box and studied its contents through cigarette smoke.

"How about number twelve, my old room?" Mark said. He tried a smile on her, with no effect. "Just for old time's sake. I'm on the eight o'clock flight out tomorrow morning."

"I know," she said and slid the number twelve key across to him.

Back at the house, as he stripped the familiar blue sheets off the bed, the scent of *J'aime* perfume rose up. He held them to his face, remembering cool blue evenings in this room. Mark put the sheets in one of the cardboard boxes and moved it to the dining room. He re-made the bed with a set of company linen Del had given him.

He stuffed all the remaining papers from the desk and dresser into the other box. At the back of the desk drawer was a wrinkled paper bag with the gold coin Cathy had bought for him. He held the heavy Mexican double-eagle in his hand and remembered the gold souk, the way Cathy had smiled and laughed, her blue eyes shining in that glittering golden light. A night full of anticipation, unclouded with plans, existing only for the moment. Mark put the coin in the box and taped it closed.

The next morning Mark ate breakfast at the mess hall and drove to the hospital to get an overdue cholera shot. He ducked inside and made his way to the inoculation area and eventually an unsmiling Egyptian nurse prepared the inoculation and gave it to him.

As he was finishing up, Mary stuck her head in and said something to the nurse. When Mark stepped out into the corridor, Mary was waiting for him, her eyes brimming with anger. She stepped up to him and tilted her deeply tanned face up at him. "Let me tell you something, Mark." They stepped to one side of the corridor. "Cathy waited until the last minute of her contract for you to make a serious commitment to her and you never did. It broke her heart."

"I'm sorry, Mary. I tried to be honest with her."

"Maybe." Mary lowered her voice. "Cathy may have seemed tough to you, but she was a rather fragile person for reasons that are none of your business. She wanted a warm and loving relationship with you, she deserved it, but you didn't provide it. She is one of the nicest people I've ever met and a wonderful pediatric nurse. You did not treat her right. You..." Mary stopped herself with some effort. "Well, doesn't matter now. Too late." She shook her head once and walked away.

At the motel Mark showered and packed his things. He took his suitcase to the living room and sat down on the rock-hard couch. He remembered the first day he'd walked into this room and looked at the Drexel couch and chair, beige with double blue stripes, at the white kitchen appliances, at the bedroom with the new Sealy mattress on it. He saw all the faces of the people he had come to know: Dale and Dick, Danny, Kurt, Brian, Jim, Ray...the card games, the days on the job site, the conviviality of the mess hall, parties at the pool, his trip to Damascus and Istanbul with Floyd and Danielle. The nights he and Cathy had made dinner here and talked about their lives in the States and the places they wanted to go, the night they had first made love in the wide bed with Roberta

Flack's "First Time Ever I saw Your Face" on the cassette player.

After a minute more, he closed the door gently and walked away.

Chapter 15

The TWA 727 slanted down over farmland, flashed past the McDonnell-Douglas hangers to a landing at St. Louis airport. Mark walked down the long concourse reveling in the accents of the people, the humid air, the Budweiser sign at a restaurant - it was great to be back. He rented a car from Avis and drove west on Highway 70 to Columbia. Allen and Carla didn't expect him until late afternoon so he'd have time to look around Columbia again before he drove to their place.

He drove down the familiar streets, parked the car at a parking meter and walked up and down Broadway for a while just window shopping.

He went into the Ninth Street Deli. The bell on the big oak door jingled and the fresh-faced girl behind the counter looked up from her book. "Hi, what can I get you?" Her Missouri accent brought a smile to his face. He ordered a barbecued pork sandwich and a Coke and took a seat near the window and watched the people walking by, the cars, the color of the summer light through the trees.

I'm home, he thought, but it still doesn't feel quite right. Like there's a plate of glass between me and all these familiar things. He squinted up at a blue sky and summer clouds. But it's not the same. Jennifer is gone and I'm not a student any more.

The waitress brought over his sandwich. "You from out of town?"

"I used to live here."

"Where do you live now?"

"Saudi Arabia. I work there."

"Wow." She stared at him without comprehension. "Must be a scary place."

"Most people there are good people. Just like any other place."

"You couldn't get me to go there. In fact, you couldn't get me on a plane at all."

"It's safer than driving a car."

She shook her head and went back to the kitchen.

Outside, he stood on the shady sidewalk watching the oak and elm trees waving in the summer wind, then he strolled to campus, across the Quadrangle, past the Engineering Building, to the Arts and Science Building. He pushed through the doors and walked down the hall, absorbing the scents and sounds of the past. Then he went down the street to the Heidelberg and drank a draft beer at the bar. A Cardinals baseball game was on TV. All around him were conversations about the gas shortages, the pizza at a new restaurant, some music he'd never heard of.

At five o'clock he bought a twelve-pack of Busch and drove out to Allen and Carla's farm. They had pork ribs on the barbecue grill and were sitting on the deck under the old elm tree back of the house. Allen, in a John Deere cap, Wrangler jeans, cowboy shirt and boots looked heavier than he remembered. His handshake was strong, his hands rough.

"Welcome back," Allen said.

Mark held up the twelve-pack, "I brought some friends." Mark passed beers around and pulled up a chair.

"You just drive down from St. Louis?" Allen asked.

"Yeah. Drove around Columbia a little. The town looks about the same."

"Oh?" Carla said. "I think it's changing pretty fast." Out across the hay field a whippoorwill called in the gathering dusk. "Well, tell us about what you've been doing, where you've been."

Mark opened another can of Busch "Went to Nepal for a couple of days. It's fantastic, the feel of Katmandu, the stillness of the mountains, the pale sunshine, you kind of feel the spirituality. Seeing the Himalayas from an airplane – spectacular."

"You guys ready to eat?" Carla had brought plates out from the kitchen.

They piled on ribs, corn on the cob, baked beans.

The radio was on, Midwestern accented ads for tire shops and discount shoe stores. Seemed familiar and strange at the same time.

Allen and Carla were talking about people Mark didn't know. He got up to fumble another beer out of the cooler and staggered.

"Jet lag," he grunted." But he drank another beer anyway while he tried to articulate some anecdotes about Saudi Arabia, but it was difficult to find things to say.

"It's amazing," Mark slurred, "The amount of

money the Saudis are putting into their military equipment, and buildings and roads, and everything, whole cities being built out of bare desert."

They didn't say anything.

"Wish I had that money," Allen said to fill the silence.

"Speaking of money," Carla said. "There's a farm nearby that's for sale." She gave Mark a look. "You keep saying you want to move back here eventually. Why not buy something now?"

"How much?"

"It's about fifteen hundred per acre. Eighty acres would be a hundred twenty thousand. On your salary that should be easy."

Mark laughed, "I only make fifteen percent more than I did right here in the States. I'm not making big money like some guys. Where is this place?"

They described the location, but he couldn't picture it. "I'll have to think about it."

Allen turned up the Cardinals game on the radio; more beers were opened. Mark took his plate to the kitchen where Carla was loading the dishwasher. He stood watching her work for a moment.

"I never meant to hurt Jennifer," he said slowly.

She turned, holding a plate. Her expression was not unsympathetic. "Well, she's doing fine. Dating an instructor at the University. They seem to be doing okay."

Carla seemed to be a great distance away. Mark set his plate down carefully on the counter and stepped back. "Well, that's good. I'm glad to hear it."

They both stood there for a moment, Mark wanted

to say I just don't love Jennifer any more. We were in love, back here, back when we were undergraduates. But, it's over!

After a minute more, Carla turned back to the dishwasher.

Mark stood there for a moment, wanting to tell her how precious the memories of the time they'd spent together was, but there was no way to say it. Instead he wandered outside and took his chair near the cooler, sipping a warm, flat beer. It's over, he said to himself, no going back. That's what I wanted and that's what I've got. She's moving ahead with her life and I'm happy for her.

The years they'd spent together flickered through his mind again, like it had so many times, but it seemed like a long time ago.

Mark took his suitcase up the narrow stair. The air smelled like old wood, fresh paint, new drapes. In his room, he peered out age-rippled glass to a dark countryside. He lay down on clean new sheets and was quickly asleep. He woke once, thinking he was in his room at Tabuk.

He'd been dreaming of cool and elegant lines on paper, black against white, architectural lines of complexity and clarity. Moonlight lay on the hardwood floor bleaching it white. On his suitcase the Saudi Customs' agents chalk marks glowed in the moonlight.

His mind took him back to the first night he had arrived in Saudi Arabia, nervously walking into the airport terminal crowded with foreigners. He thought about his room at the IIAA motel in Tabuk and the first night he and Cathy had made love there. He thought about the house they had shared, and the trips they had made to Riyadh to Aqaba, and out into the silent desert. He looked back on the trip he had made with Dale and Tom to Bangkok, and to the Himalayas in dawn light on his trip to Nepal.

He got up and opened the window a little wider, letting the soft night air into the room. Summer thunder rumbled far away and there was a flicker of distant lightning. On the deck below, three empty chairs still sat in a little circle around the grill. It had always been the four of them over the years, doing things together, Allen and Carla, Jennifer and him. Now those days were gone. He lay back down and watched clouds moving across the moon in absolute silence.

He woke well rested. Outside, the lawn and the fields glowed gold and green.

Breakfast was full of jokes and energy, plans for the day. "I got your work schedule all laid out for you," Allen joked. "Got to cut and rake hay, then get it baled."

"I need Mark to help me at my store this morning," Carla said, "Stacking boxes, cleaning up."

"Then we need to get the skid loader hydraulics repaired," Allen said. "That'll take you a couple of hours."

The phone rang. "It's for you," Allen told Mark.

It was the IIAA office in Washington D.C. "The Saudi's have given us a big new supplemental agreement. A hundred million dollars," Hamilton, IIAA's personnel manager in Washington D.C. told him. "We need everybody we've got back in-country ASAP. How soon can you get back?"

Mark started to say, 'I just got here,' when he realized he wouldn't mind going back. After the phone call he went back to his bacon and eggs and biscuits. "Big emergency?" Allen asked.

"Well, a big new job anyway. They need everybody back in-country right away."

"Afraid I'll put you to work here on the farm, right?" Allen said. He checked his watch. "I've got to get over to Don's place. I told him I'd help him with the hay this morning, so..."

They shook hands; Mark hugged Carla. "Come back for a visit soon," she said. "I've got to get to the store. Just pull the door closed behind you when you leave. Think about that property for sale!"

She drove off.

Mark stood outside on the deck watching the humid wind swaying the oak and walnut trees, the morning clouds meeting the green distances, woods and hay fields. He went inside and sat down in a wooden rocking chair in the living room and looked around the room for a few moments. Then he got his suitcase, put it in the car and stood there for a moment. A grasshopper leaped in the dusty silence, humidity shimmered the green hills out at the flat horizon, the weeds were dusty near the gravel driveway. He squinted up toward the sun, "I guess it's impossible to communicate all the things I've seen and done."

At the airport, waiting for his flight, he found a seat facing a bit of lawn outside the glass windows, trying to prolong his immersion in the familiar Midwest. He waited for the nostalgic familiarity to wash over him, but the blue sky and the sound of lawn mowers, the cracked pavement and the muddy cars still seemed very remote.

After a while he bought a magazine and found a seat near his gate until his flight was called. It would be good to get back. Back to work, back to the familiar challenging routines of construction management, the endless desert horizon, the empty white sky.

Afterword

Mark Exner returned to Saudi Arabia with IIAA and was given a promotion to project engineer on the Riyadh General Hospital expansion. On his first vacation back to Missouri he made a down payment on a small piece of property not far from Allen and Carla Hayes' farm.

Cathy Locke married an accountant from Ocean City, Maryland three months after she returned to the United States from Tabuk. She happily gave up her nursing career and became a full-time mother. She is widely admired among their friends for homemaking, especially her cooking skills, and still serves very large portions on very small plates.

Jennifer Campion completed a masters' degree program in Information Science at the University of Missouri where she met and married an associate professor of History. They moved to the Kansas suburbs of Kansas City where he is on the tenure track at the University of Missouri at Kansas City. They live comfortably, travel to Europe occasionally, and sometimes talk about retiring to a place deep in the country.

Dale Parker returned to Nebraska after one year in Saudi Arabia and never went overseas again. He bought the fitness club he had previously managed, expanded it, invested in three more clubs, and a variety of real estate ventures in central Nebraska, all quite successful. He married his high school sweetheart and they live in a ten thousand square foot house on the fairway at Silverlake Country Club in North Platte.

After Tom Farris and Trish were divorced, Tom returned to Long Beach, California and took a position as sales rep for Lexus of Long Beach. He remarried and was again divorced within a year. He still likes to entertain, and often regales his friends with stories of Saudi Arabia and Thailand which grow more dramatic as time passes.

Jim Redding resigned from IIAA when the Tabuk office closed and took a job with a Korean construction company working for IIAA in Saudi Arabia. He did an exceptionally good job at this because of his familiarity with IIAA procedures. He was paid well and saved almost all of it. At age sixty five, Jim and Ti retired to Rancho Cordova near Sacramento, California and paid cash for a house. At age sixty-seven, Jim was diagnosed with lung cancer and died six months later. He willed the house, now worth one and a half million dollars, to Ti.

Floyd and Danielle Calvin returned to Florida, where Floyd worked in the Operations division of the Corps of Engineers until his retirement a year later when they moved to a modest retirement home in Palatka, Florida. Not long afterward, Danielle came out of Publix and found Floyd slumped over the wheel of their Cadillac, dead of a heart attack. A month later, Danielle sold the house and returned to the small village in France where she had grown up.

Brian and Linda Zeller were divorced; Linda retained custody of their children. Brian took an administrative job with the Pennsylvania Department of Transportation. He tried to get another assignment with IIAA in Saudi Arabia, but was not offered a position. Ray Barton took a staff position with the Ohio department of Health Services in Dayton. Danny Hager returned to Fort Worth and took a job with the Texas Department of Corrections, facilities management department. Ed Preece moved to Jackson Mississippi and spent a year working for the Mississippi River Commission before retiring. Andy Petri returned to Omaha and was killed in a car accident three months later. Alcohol was involved.

Kurt Hess left Kendall International and worked as a project manager for Caraway Construction in Vancouver, British Columbia, Canada. He became a Canadian citizen and formed his own consulting firm, which prospered in the Vancouver building boom. Tony Cross went back to Britain, took a job with the J.D. Laing Company as quality control manager and was assigned to the airport expansion in Dubai. Dick Davis left Tabuk and spent the next six months in Honolulu drinking and carousing with his buddies at the Down Under Bar in Ala Moana. He applied for and was accepted for a two-year assignment with Intercontinent Engineering at Anderson Airbase, Guam, but before departing went on a fishing trip with two friends. They returned to Ala Wai yacht basin, Dick waved goodbye, walked away toward his condo in Waikiki, and was never heard from again.

www.ingramcontent.com/pod-product-compliance
Lightning Source LLC
Chambersburg PA
CBHW051816090426
42736CB00011B/1502